Naming Infinity

Naming Infinity

A True Story of Religious Mysticism and Mathematical Creativity

LOREN GRAHAM *and*
JEAN-MICHEL KANTOR

THE BELKNAP PRESS OF
HARVARD UNIVERSITY PRESS
Cambridge, Massachusetts
London, England
2009

Copyright © 2009 by the President and Fellows of Harvard College
All rights reserved
Printed in the United States of America

Library of Congress Cataloging-in-Publication Data
Graham, Loren R.

Naming infinity : a true story of religious mysticism and mathematical creativity / Loren Graham and Jean-Michel Kantor.

    p. cm.

Includes bibliographical references and index.

ISBN 978-0-674-03293-4 (alk. paper)

1. Mathematics—Russia (Federation)—Religious aspects.
2. Mysticism—Russia (Federation)
3. Mathematics—Russia (Federation)—Philosophy.
4. Mathematics—France—Religious aspects.
5. Mathematics—France—Philosophy. 6. Set theory.
I. Kantor, Jean-Michel. II. Title.

QA27.R8G73   2009

510.947′0904—dc22   2008041334

CONTENTS

Introduction	1
1. Storming a Monastery	7
2. A Crisis in Mathematics	19
3. The French Trio: Borel, Lebesgue, Baire	33
4. The Russian Trio: Egorov, Luzin, Florensky	66
5. Russian Mathematics and Mysticism	91
6. The Legendary Lusitania	101
7. Fates of the Russian Trio	125
8. Lusitania and After	162
9. The Human in Mathematics, Then and Now	188
Appendix: Luzin's Personal Archives	205
Notes	212
Acknowledgments	228
Index	231

ILLUSTRATIONS

Framed photos of Dmitri Egorov and Pavel Florensky. Photographed by Loren Graham in the basement of the Church of St. Tatiana the Martyr, 2004.   4

Monastery of St. Pantaleimon, Mt. Athos, Greece.   8

Larger and larger circles with segment approaching straight line, as suggested by Nicholas of Cusa.   25

Cantor ternary set.   27

Émile Borel. Reproduced by permission of Institut Mittag-Leffler and *Acta Mathematica*.   44

Henri Poincaré. Reproduced by permission of Institut Mittag-Leffler and *Acta Mathematica*.   46

Henri Lebesgue. Reproduced by permission of *L'enseignement mathématique*.   48

René Baire. Reproduced by permission of Institut Mittag-Leffler and *Acta Mathematica*.   52

Arnaud Denjoy.   54

Jacques Hadamard. Reproduced by permission of Institut Mittag-Leffler and *Acta Mathematica*. 57

Charles-Émile Picard. Reproduced by permission of Institut Mittag-Leffler and *Acta Mathematica*. 60

Hotel Parisiana on the rue Tournefort in Paris, c. 1915. Reproduced from Anna Radwan, *Mémoire des rues* (Paris: Parimagine, 2006), p. 111. 81

Nikolai Luzin in 1917. Courtesy of *Uspekhi matematicheskikh nauk*. 85

Pavel Florensky. From Charles E. Ford, "Dmitrii Egorov: Mathematics and Religion in Moscow," *The Mathematical Intelligencer*, 13 (1991), pp. 24–30. Reproduced with the kind permission of Springer Science and Business Media. 89

Building of the old Moscow State University where the Lusitania seminars were held. Photograph by Loren Graham. 104

Luzin's apartment on Arbat Street, Moscow. Photograph by Loren Graham. 107

Interior of Church of St. Tatiana the Martyr, Moscow. Photograph by Loren Graham. 111

Nikolai Luzin, Waclaw Sierpinski, and Dmitri Egorov in Egorov's apartment in Moscow. Photograph courtesy of N. S. Ermolaeva and Springer Science and Business Media. 120

Otto Shmidt. Courtesy of the Shmidt Institute of Physics of the Earth, Russian Academy of Sciences, Moscow. http://www.ifz.ru/schmidt.html. 123

"A Temple of the Machine-Worshippers." Drawing by Vladimir Krinski, c. 1925. 128

Illustrations

Ernst Kol'man. Reproduced with the permission of Chalidze Publications from Ernst Kol'man, *My ne dolzhny byli tak zhit'* (New York: Chalidze Publications, 1982).   130

Nikolai Chebotaryov. Courtesy of the State University of Kazan, the Museum of History.   133

Hospital in Kazan where Maria Smirnitskaia tried to save Egorov. Photograph by Loren Graham, 2004.   137

Dmitri Egorov's gravestone, Arskoe Cemetery, Kazan. Photograph by Loren Graham, 2004.   139

Nina Bari. Courtesy of Douglas Ewan Cameron, from his collection of pictures in the history of mathematics and *Uspekhi matematicheskikh nauk*.   154

The Luzins with the Denjoy family on the island of Oléron, Brittany. Courtesy of N. S. Ermolaeva.   156

Peter Kapitsa. Courtesy of the Institute of the History of Science and Technology, Academy of Sciences, Moscow, and Sergei Kapitsa.   159

Genealogical chart of the Moscow School of Mathematics.   163

Ludmila Keldysh. Courtesy of A. Chernavsky, "Ljudmila Vsevolodovna Keldysh (to her centenary)," *Newsletter of the European Mathematical Society*, 58 (December 2005), p. 27.   165

Lev Shnirel'man. Courtesy of *Uspekhi matematicheskikh nauk*.   168

Pavel Alexandrov, L. E. J. Brouwer, and Pavel Uryson in Amsterdam, 1924. Courtesy of Douglas Ewan Cameron, from his collection of pictures in the history of mathematics.   176

Grave of Pavel Uryson (Urysohn) at Batz-sur-Mer, France. Photograph by Jean-Michel Kantor. 178

Pavel Alexandrov. Courtesy of Douglas Ewan Cameron, from his collection of pictures in the history of mathematics. 179

Andrei Kolmogorov. Courtesy of *Uspekhi matematicheskikh nauk*. 181

Pavel Alexandrov swimming. Courtesy of Douglas Ewan Cameron, from his collection of pictures in the history of mathematics. 183

Alexandrov and Kolmogorov together. Courtesy of Douglas Ewan Cameron, from his collection of pictures in the history of mathematics. 185

Naming Infinity

Introduction

IN THE SUMMER OF 2004 Loren Graham was invited to the Moscow apartment of a prominent mathematician known to be in sympathy with a religious belief called "Name Worshipping" that had been labeled a heresy by the Russian Orthodox Church. The mathematician implied he was a Name Worshipper without stating it outright, and he intimated that this religious heresy had something to do with mathematics.

Graham had sought out the Russian scientist at the suggestion of a French mathematician, Jean-Michel Kantor, with whom he had begun discussions of religion and mathematics three years earlier. Graham, an American historian of science, had long known that there was an interesting unexplored story about the beginnings of the famed Moscow School of Mathematics early in the twentieth century. After reading a book by Graham that hinted at this story, Kantor immediately contacted Graham to tell him that he knew something about these events. The two met in 2002 and found, to their mutual excitement, that their respective pieces of the narrative had many things in common. Moreover, Kantor told Graham that the story was not just about Russian mathematicians, but about French and world mathematics as well. As Kantor put it, in the early years of the twentieth century mathematics had fallen into such strong contradictions that it was very difficult for mathematicians to see how to go forward. The French, leading in the field, and the Russians, trying to catch up, took two different approaches to the same problems. The French had mixed feelings about the issues; they engaged in passionate discussions, and important breakthroughs were made by Émile Borel, René Baire, and Henri Lebesgue, but they ended up sticking to their rationalistic, Cartesian presuppositions. The Russians, learning the new mathematics from the Paris seminars they attended, were stimulated by mystical and intuitional approaches connected to a religious heresy, Name Worshipping, to which several of them were loyal.

The two of us began digging more deeply into the story, reading everything we could find about the beginnings of set theory in France and Name Worshipping in Russia, and looking for people in both countries who could tell us more. The trail led to the Russian mathematician in Moscow who agreed to talk to Graham about Name Worshipping.

The mathematician's apartment was a typical one built in Soviet times—small and cramped, with just enough space to live and work. The hallway connecting the apartment's four rooms was lined with bookcases filled with works on mathematics, linguistics, philosophy, theology, and rare books on Name Worshipping. In one of the few empty wall spaces hung framed photographs of two men who, according to the mathematician, were early leaders of Name Worshipping: Professor Dmitri Egorov and Father Pavel Florensky. Another photograph showed the Pantaleimon Monastery on Mt. Athos in Greece, which the mathematician asserted was the early home of Name Worshipping. Yet another photo displayed a book cover with the title "Philosophy of the Name," written by a Russian philosopher who had subscribed to Name Worshipping in the 1920s.

Graham asked if it would be possible to witness a Name Worshipper in the Jesus Prayer trance, which he had recently learned was at the center of the Name Worshipping faith. "No," replied the mathematician, "this practice is very intimate, and is best done alone. For

Introduction

you to witness it would be considered an intrusion. However, if you are looking for some evidence of Name Worshipping today I would suggest that you visit the basement of the Church of St. Tatiana the Martyr. In that basement is a spot that has recently become sacred to Name Worshippers."

Graham knew about this church; decades earlier it had been closed down during an anti-religious campaign by Soviet authorities and converted into a student club and theater. Now, in the post-Soviet period, it has been restored as the official church of Moscow University, as it was before the Russian Revolution. It is located on the old campus near the Kremlin, in a building attached to the one that housed the Department of Mathematics in the heyday of Dmitri Egorov and Nikolai Luzin, founders of the Moscow School of Mathematics. It is the church where they often went to pray. Graham asked the mathematician, "When I go into the basement, how will I know when I have reached the sacred spot?" The mathematician replied, "You will know when you get there."

What was the connection between Name Worshipping and mathematics? And why did the mathematician speak of Name Worshipping in such a cautious way? The next day Graham went to the Church of St. Tatiana the Martyr and made his way to the underground level with its whitewashed walls, where at first he found nothing of note. Then he saw an alcove, leading to a corner where the walls came together at less than the normal ninety degrees, and there he found pictures of the same two men whose faces adorned the apartment of the mathematician: Dmitri Egorov, longtime president of the Moscow Mathematical Society, and Pavel Florensky, his former student, who became both a scientist and an Orthodox priest. Graham was standing in the place where Name Worshippers came to practice the Jesus Prayer.

Just after taking photographs of the two portraits, Graham heard steps behind him and turned to see a young man with a disapproving look on his face. The man came up to him and warned, "Vam nado uiti" ("You must leave"). Graham sensed the same intrusion into a

Framed photos of Dmitri Egorov and Pavel Florensky, photographed by Loren Graham in the basement of the Church of St. Tatiana the Martyr, 2004.

mystery as he had when the Russian mathematician rejected his request to see a Name Worshipper in the Jesus Prayer trance. He put away his camera and left. Who was this young man? A Name Worshipper? An employee of the church? He was not wearing clerical robes, and he looked as if he might have been a student. Caught up in the story that was beginning to unfold, Graham hoped that he was a talented young mathematician.

The two of us continued our research on the French school of mathematics and Name Worshipping, working in French and Russian libraries and archives. In December 2004, during a research trip in Moscow, Graham felt drawn once again to the basement of the Church of St. Tatiana the Martyr and its connection to Name Worshipping. He made his way down to the basement and found, to his surprise, that it was a completely different place. The sacred spot had been eliminated by the Church, which had finally realized that Name Worshippers were coming to the basement to celebrate their heresy, one condemned by the official Russian Orthodox Church. Now a

Introduction

regular chapel of the Church occupies the basement, with a priest watching over it and ensuring the orthodoxy of all worshippers. Jesus Prayer trances are no longer practiced there. Thus, the struggle over Name Worshipping continues today. Although they agree on little else, the Communists and the Church officials both oppose it.

This book is devoted to a little known but exemplary episode in the recent history of the relationship of mathematics and religion, all within the context of much larger issues of religious heresy, rational thought, politics, and science. It is intended for general readers, although we hope that mathematicians will also find it worthwhile. It is the story of an initial breakthrough by a German mathematician that was picked up and developed further by the French, who eventually stalled, but who taught the new developments to Russian mathematicians; the Russians then returned to their homeland and managed to push onward to a fundamental insight.

At the center of the story is an encounter at the beginning of the twentieth century between mathematicians working on set theory and the religious practices of the heretical Name Worshippers in Russia. Set theory was at first brilliantly developed in France but then underwent a profound crisis, only to have the Russians enter the scene with a new energy. We will describe how two different states of mind connected with two different cultural contexts led to contrasting results: French skepticism and hesitation, Russian creativity and advancement. A central idea of this book is that a religious heresy was instrumental in helping the birth of a new field of modern mathematics.

The originality of Russian mathematics blossomed in the early twentieth century, when Dmitri Egorov, Nikolai Luzin, and their students developed a very specific approach to the new set theory which was already the center of polemics for many European mathematicians and philosophers. Egorov's and Luzin's achievements have up to now attracted relatively little attention from the public or historians of science, even though the work of the Moscow School of Mathematics, which they founded, is well known to professional mathematicians. What is not known is that their work was linked with intense mysticism, political persecution, and personal drama. It is this story that we will tell here—a story that sheds light on the creative process of mathematics itself.

1

Storming a Monastery

> "Heretic, crocodile from the sea, seven-eyed serpent, wolf in sheep's clothing!"
> —*Description by a Mt. Athos monk of a church official sent from St. Petersburg to subdue him and his colleagues*

IN EARLY JUNE, 1913, several ships from the Imperial Russian Navy, acting on Tsar Nicholas II's orders, steamed into the azure waters surrounding the holy site of Mt. Athos in Greece, a center of Orthodox Christianity for a thousand years. The ships, the gunboat *Donets* and the transport ships *Tsar* and *Kherson*, anchored near the Pantaleimon Monastery, a sacred bedrock of Russian Orthodoxy and residence of hundreds of Russian monks. On board the *Tsar* were 118 marines under the command of Z. A. Shipulinsky and four other officers.

On June 13, Shipulinsky ordered that the monastery be stormed. The heavily armed marines made their way in small boats to the monastery dock, where the men disembarked. They then proceeded to the largest space of the religious complex, the Pokrovsky Cathedral, which at that moment was nearly empty. There Shipulinsky met with several of the religious ascetics and told them that they were to inform all their brethren to leave their cells and assemble in the cathedral. When the monks learned of the order, they refused, barri-

Monastery of St. Pantaleimon, Mt. Athos, Greece.

cading the doors of their cells with furniture and boards. Inside they fell on their knees and began crying "Lord, Have Mercy!" *(Gospodi pomilui)*, and many of them launched into a unique prayer, one that was causing controversy in the Church, called "The Jesus Prayer."

It was because of this prayer that the Russian marines were here to begin with. The practice of the prayer, called heretical by some leaders of the Russian Orthodox Church, had been causing great disorder on Mt. Athos. This peninsula in the Aegean Sea had been the location of Orthodox monasteries since the early Christian era, and Russians were among the most numerous of the monks, with several thousand usually present. For centuries the Ottoman Turks had occupied most of the Balkans, including Athos, but they granted the monks there near-autonomy, allowing them to do what they wanted so long as they did not directly challenge the Turks. The Russian monks on Athos usually looked to their homeland government in St. Petersburg for support and protection, but the collapse of the Ottoman Empire and the retreat of the Turks from Athos in 1912 led to a delicate situation. Would the autonomy of the Holy Mountain and the Russian influence continue under Greek rule? The Greeks, who shared the Orthodox faith of the Russians, seemed ready to grant the monasteries considerable freedom and withdrew their soldiers. The Russian monks then began to call for the creation of an independent republic of Athos that would amount to a protectorate of the tsarist government—a challenge to the Greeks.

In the middle of this diplomatic problem, a theological dispute erupted which unnerved the Russian governmental and clerical leaders. The last thing the Church and government in St. Petersburg needed was a bunch of monks fighting one another over a prayer, giving the Greeks a pretext for intervention and elimination of the traditional autonomy of Mt. Athos.

A dramatic fight was indeed going on among the monks between those who supported the practice of the Jesus Prayer (known as Name Worshippers) and those who did not (the Anti-Name Worshippers). The struggle often took its sharpest form when administrative leaders of the monasteries were being chosen: each side wanted its own people to lead. The acrimony increased rapidly, with actual physical conflicts; each side tried to eject the members of the other camp from the monasteries, and sometimes succeeded, at least temporarily. In several instances monks were thrown or jumped out of windows during scuffles. Each side declared that the other was no longer eligible for communion. Each side appealed to higher authorities for support—to the Russian consul in Salonika, to the Russian ambassador in Constantinople, to the Holy Synod in St. Petersburg, and, eventually, to the tsar himself. Word spread throughout the Balkans and the Russian Empire that "disorders" were rife in the monasteries at the Holy Mountain of Athos.

At first the Russian government tried to subdue the rebellious monks by nonviolent means. In February 1913 a blockade was imposed on the Name Worshipping monks on Mt. Athos, whose stronghold was the Pantaleimon Monastery. That monastery was deprived of food supplies, financial support, and postal service for five months. However, the stubborn monks proved resourceful in obtaining what they needed through local contacts with Greek peasants and sympathetic monks in neighboring monasteries. The stories of "revolts" and "mutinies" among the monks continued, and eventually the Greek government responded by saying to the leaders of the monasteries, in effect, "Bring order to the monasteries yourselves or we will do it for you." Greek troops assembled nearby in preparation for occupying the monasteries if necessary.

This international difficulty goes a long way toward explaining why the tsarist government yielded to the plea of the top leaders of the Russian Orthodox Church to suppress the Name Worshippers at Mt. Athos with military force. Tsar Nicholas II was not particularly interested in the theological dispute, and his wife Alexandra was even sympathetic to the Name Worshippers, but his advisers, especially V. K. Sabler, the head of the Holy Synod, told him that if the disorder at Mt. Athos continued not only would the Russian Orthodox faith be hopelessly split by schism, but the Russian government would lose much of its influence in a crucial area of Greece and the Balkans. Faced with this opinion, Nicholas reluctantly agreed to the invasion of the monastery.

Following the tsar's orders, Officer Shipulinsky led his marines into the monastery cathedral and demanded that the monks come out of their cells and assemble before him. When he was ignored, he ordered his men to prepare for conflict. The marines unrolled high-pressure water cannons and also set up several machine guns. They then tore down the barricades at the entrances to the monastic cells and aimed the water cannons at the men inside. What happened next is still today, almost a century later, hotly disputed. Sources sympathetic to the monks say that the marines opened fire, killing four of the recluses and wounding forty-eight others. The official Russian navy accounts say that the marines were met with "criminal resistance" requiring force to overcome, but maintain that no one was killed even though some "fanatics" were wounded. Certainly it was a bloody affair; the marines beat the monks with their bayonets and rifle butts and bashed many heads.

The marines flushed the recluses from their cells and herded them into the cathedral. There the officer announced to the soaked, terrified, and injured monks that they must either renounce their heretical beliefs or be arrested. Archbishop Nikon of Vologda accompanied the marines; a representative of the highest authorities in the Russian Orthodox Church, he lectured the assembled monks on the details of their "Name Worshipping heresy" in a voice trembling with fear and emotion: "You mistakenly believe that names are the same as God. But I tell you that names, even of divine beings, are not God themselves. The name of Jesus is not God. And the Son is less than the Father. Even Jesus said, 'the Father is greater than me.' But you believe you possess both Christ and God." Some of the monks responded by crying out that the archbishop and the marines represented the "Anti-Christ." Many of them shouted what had become an unofficial slogan of the Name Worshippers: "The Name of God *is* God" ("Imia Bozhie *est'* sam Bog"). One monk called Nikon a "heretic, crocodile from the sea, seven-eyed serpent, wolf in sheep's clothing." Nikon angrily pounded his ornamented crosier on the floor and demanded that the assembled monks be polled individually, stating whether they renounced their heresy or remained obstinate.[1]

According to the official count, 661 monks stated that they did not support the doctrine of "Name Worshipping," but 517 were adamant and declared that they were, and would remain, "Name Worshippers." Another 360 refused to participate in the poll and were considered by the archbishop to be on the side of the heretics. Several dozen others were so badly injured that they were taken away for medical care and not polled. In the nearby Andreevsky Monastery and elsewhere on Mt. Athos the archbishop found other Russian monks whom he considered to be unrepentant Name Worshippers. Sobered by the violence in the Pantaleimon Monastery, they did not resist arrest. Eventually approximately a thousand monks were taken back to Russia under detention, most of them on a ship converted into a prison, the *Kherson*, but others on the steamship *Chikhachev*.

When on July 13 and 14 the *Kherson* and the *Chikhachev* arrived in Odessa, a major Ukrainian/Russian port on the Black Sea, the tsarist police there interrogated the imprisoned monks and then divided them into groups. Some were so old and feeble that they were permitted to go to local monasteries that might care for them; eight were returned to Athos; and forty were accused of criminal activity and sent to prisons. The rest—eight hundred or so—were defrocked and told that they could not return to Mt. Athos or reside in the cities of St. Petersburg or Moscow. Instead, according to the Russian governmental system of assigned residence (*propiska*), they were exiled to provincial and rural locations all over Russia. Sometimes they ended up in remote monasteries whose administrators did not recognize, or perhaps did not know about, their defrocked status. One of the leaders, Alexander Bulatovich (known as monk Antony), was sent to his family estate near Kharkov, where he was joined by many of his comrades.

The doctrine that inspired this passion, Name Worshipping, has roots going back almost to the beginning of Christianity; its twentieth-century chapter began in 1907, when a monk and *starets* (respected elder) named Ilarion published a book entitled *On the Mountains of the Caucasus*. In this text Ilarion told of his mystical and spiritual experiences when reciting the Jesus Prayer, an established prayer in the Orthodox tradition, but one to which Ilarion gave a special significance.

Between 1872 and 1892 (the dates are only approximately known) Ilarion was a monk at the Pantaleimon Monastery on Mt. Athos, where he made use of the extensive library to immerse himself in the history of Christian mystics, going back to the fourth century. At that time hermits in the deserts of Palestine developed a new practice of prayer, the "Prayer of the Heart," intended to obtain quietness by physical and mental fusion with God (they would later be called hesychast monks; *hesychia* means rest or stillness). The prayer techniques combined hundreds of repetitions of short sequences of the same words (glossalia, "praying without ceasing," a quotation from the apostle Paul) with control of breathing and the heartbeat. Under attack from the intellectual, more rationalist monks of Byzantium around 1340, the hesychast monks were supported by Grigorii

Palama (1296–1359), who was already practicing the Prayer of Jesus on Mt. Athos.

Around 1892 Ilarion left Athos and went to a mountain monastery in the Caucasus (in Abkhazia, currently a breakaway province of Georgia), seeking more isolation and peace. There he wrote his book, copies of which he sent back to his former colleagues at the Pantaleimon Monastery, where it became very popular.

Ilarion believed that he made contact with God by chanting the names of Christ and God over and over again until his whole body reached a state of religious ecstasy in which even the beating of his heart, in addition to his breathing cycle, was supposedly in tune with the chanted words "Christ" and "God."

This state of ecstasy and insight was vividly described by J. D. Salinger in his 1961 novel *Franny and Zooey*. Salinger has Franny observing:

> If you keep saying that prayer [the Jesus Prayer] over and over again—you only have to just do it with your *lips* at first—then eventually what happens, the prayer becomes self-active. Something *happens* after a while. I don't know what, but something happens, and the words get synchronized with the person's heartbeats, and then you're actually praying without ceasing. Which has a really tremendous, mystical effect on your whole outlook. I mean that's the whole *point* of it, more or less. I mean you do it to purify your whole outlook and get an absolutely new conception of what everything's about.[2]

The words which usually formed the heart of the prayer were "Gospodi Iisuse Khriste, Syne Bozhii, pomilui mia greshnago" ("Lord Jesus Christ, Son of God, have mercy on me, a sinner"). However, adepts at the prayer often shortened these eight words to just three, "Gospodi Iisuse Khriste" ("Lord Jesus Christ") or even just one, "Iisuse" ("Jesus").

According to Ilarion, learning to recite the Jesus Prayer in the right way was a process requiring much practice that could last for years. The communion with God that the prayer allegedly brought involved three stages of immersion. First was the "oral prayer," in which the spoken names of God and Jesus were the main concern of the worshipper. Then, if the person praying was sufficiently devout and concentrated on the task, the prayer could move to the "mental" stage, when "the mind starts to cling to the words of the prayer, seeing in them the Lord's presence." Last came the "Prayer of the Heart," when the heart gains "spiritual élan" and "illumination" and the person achieves a "oneness" with God.[3]

Ilarion warned practitioners not to rush the process of achieving these various stages, but to allow the process to follow its own tempo. If the person praying tries to hasten the final stage, warm blood descends to lower parts of the body, according to Ilarion, and can even lead to "sexual arousal." Thus the practitioner of the Jesus Prayer was dealing with a process that if done right, its adherents maintained, brought humans into the closest possible contact with God, but if done incorrectly, could lead to sin. This challenge and temptation may help explain why the licentious and notorious Rasputin, who claimed to have healing powers and who was adviser to the tsarina Alexandra, became a supporter of Name Worshipping.

The arrival of Ilarion's book at Mt. Athos in 1907 at first attracted only mild favorable attention. Ilarion was obviously a devout Christian, and the monks in the Pantaleimon Monastery were particularly interested in the book because many of them remembered Ilarion personally from the time he had been there. One can easily imagine the monks in their cells in the monastery practicing the Jesus Prayer, sometimes getting it right, sometimes getting it wrong. Eventually more and more monks read the book, and its influence began to spread. The Patriarch of Constantinople, Ioakim III, spoke positively of the book. (He probably looked at it only briefly; later he and his advisers would study it more carefully and condemn it.)

One priest at the Pantaleimon Monastery, Father Khrisanf, reviewed the book in 1909 and objected strenuously to its contents. According to Khrisanf, the Name Worshippers made a fundamental theological error in equating God or Jesus with their names. By locating the essence of God outside of Himself, in His name, said Khrisanf, the Name Worshippers were falling prey to the heresy of pantheism (he evidently meant to say "polytheism," the belief that there is more than one divinity). After all, how many different names have been given to God in different creeds?

In the months after the storming of Athos, the tsar wavered about the correctness of his decision. In February 1914 he actually gave a hearing to several of the Name Worshippers, who unsuccessfully pleaded for a pardon. In May 1914 the Metropolitan of Moscow, Makarii, decided to allow Name Worshippers to participate in church services. (Tsar Nicholas had recently written him a letter urging lenience.)[4] The Holy Synod, the highest authority in the Church, in the same month ruled that Name Worshippers could stay within the church even though their theological beliefs were still considered to be a heresy. But the exact status of Name Worshipping within Russian Orthodoxy remained unclear. In September 1917 the Church assembled a council (Pomestny Sobor) to consider the question, and strong arguments were given both for and against Name Worshipping. One of the defenders of the cult was Pavel Florensky, a former mathematician just ordained as a priest. (Florensky's personal papers show that his support of the Mt. Athos Name Worshippers began as early as March 1913, before the armed seizure of the Pantaleimon Monastery.) Florensky was a friend of several of the leading mathematicians in Russia and would play an important role in subsequent mathematical events. The Church council failed to come to an agreement on Name Worshipping, and its work was aborted by the October 1917 Revolution, which brought the Communists to power.

After being dispersed from Odessa, most of the Name Worshippers remained committed to their form of faith and did not consider themselves defeated. In rural Russia they continued their practices, now officially defined as "heretical," and gradually enlarged their following. The advent of World War I in less than a year meant that the tsarist government had more pressing concerns than quarreling with religious dissidents. Name Worshipping silently grew in strength. The seizure of power by the Communists in 1917 actually at first gave the Name Worshippers an unusual opportunity (later they were cruelly persecuted). They were already an underground faith, banned from the official churches, so when the Soviet regime launched an anti-religious campaign and began to close churches and arrest priests, the Name Worshippers, being invisible, were initially not repressed. Indeed, they seemed to prosper from their banned status. They needed no church buildings, no priests, no church administration. Furthermore, they were not tainted with the compromises with the Soviet authorities that the official Orthodox Church was soon forced to make in order to exist. Reciting the Jesus Prayer and worshipping the names of Christ and God were practices that one could do alone, in one's study or even in a closet ("When thou prayest, enter into thy closet"; Matthew 6:6).

Gradually the Name Worshippers began to extend their influence to the cities where their presence was forbidden. A movement that in its first stages was often supported by uneducated, sometimes even illiterate, monks gradually attracted the attention of the urban intelligentsia, especially mathematicians and philosophers. René Fülöp-Miller, a central European journalist who visited Moscow in the early 1920s, became intrigued with the underground religious movement of Name Worshipping. He interviewed some of its adherents and announced, "The best men of Russia lead this school, which proclaims the magic power of the divine name; it is from the spread of its religious doctrines that the true revival of Russian religion is generally expected."[5] One of those "best men" was Dmitri Egorov, professor of mathematics at Moscow University and future president of the Moscow Mathematical Society.

The uncertainty of the official Church about Name Worshipping persisted. In October 1918, the Holy Synod reversed itself again and said that Name Worshippers could not participate in Church services unless they repented—something that many Name Worshippers would not do. In January 1919 their *de facto* leader, Alexander (Antony) Bulatovich, recently returned from the war in which he had served as a priest in the Russian army, gave up his effort to persuade the Church to recognize Name Worshipping and returned to his family estate near Kharkov. There on the night of December 5–6, 1919, he was murdered in mysterious circumstances. Some people said that he was killed by robbers; others said the intruders were Red Army soldiers.

After Bulatovich's death, a leadership role among the followers of the movement (Name Worshippers had no formal hierarchy) was taken by the archimandrite David, who still hoped to win over the support of the mainstream Orthodox Church. In the early twenties David established a "Name Worshippers' Circle" in Moscow at about the same time that other such circles were arising elsewhere in Russia.[6] He tried to involve in his circle priests and high officials of the Church who had not previously affiliated themselves with the Name Worshipper cult. He even, several times, managed to participate in religious services with the patriarch himself, Tikhon. One of the members of a Name Worshipping circle was the mathematician Dmitri Egorov, who actually met with the patriarch of the Church and begged him to forgive the Name Worshippers.[7]

Some of the other Name Worshipping groups had already given up on gaining the support of the Church hierarchy, and perhaps even took pride in their "heretical" status. Thus there were differences in emphasis among the Name Worshippers. David, still reaching out to the top leaders of the Church, went to considerable lengths to try to prove that the charge that the Name Worshippers were cultists believing in magic or polytheism was incorrect. He agreed that if one took literally the assertion "The Name of God *is* God," the charge of polytheism had some plausibility, since the name of God is different in many different languages. He tried to avoid this conclusion by saying that the name of God should not be understood in terms of "letters" or "specific words," but instead as an "essence" that stands behind the name. David criticized some of his fellow Name Worshippers, including the priest Pavel Florensky, for believing in the "magic" of the word "God" ("Bog").

Florensky refuted the charge that he believed literally in the divinity of the letters making up the word "God." He enlisted the help of philosophers and writers like Aleksei Losev and Sergei Bulgakov in giving his Name Worshipper circle a Neoplatonistic orientation that tried to reconcile his ideas with Christianity. He and his supporters were not interested in trying to win over the official Church; rather, they were devoted to exploring the meaning of symbolism, linguistics, and "signifiers," of which they saw the word "God" as the most important. Florensky, who mastered many languages and had a deep knowledge of Patristic traditions, managed to combine a deeply intellectual interpretation of Name Worshipping with a mystical, even magical, view of the universe. He was sharply critical of what he called "Western positivism"; instead he favored traditional Orthodox mysticism while opposing, at the same time, the interpretation of that mysticism which the Church leadership tried to impose. Partly because of this intellectualist turn of the Florensky-Losev group, followers of much more recent developments, such as the theories of Jacques Derrida, have displayed interest in Florensky and Losev.[8]

Florensky was particularly devoted to the relevance of Name Worshipping to mathematics, the field in which he was trained at Moscow University by Dmitri Egorov. Florensky saw a relationship between the naming of "God" and the naming of sets in set theory: both God and sets were made real by their naming. In fact, the "set of all sets" might be God Himself.

2

A Crisis in Mathematics

> ... The doctrine of sets, the *Mengenlehre*, which postulates and explores the vast numbers that an immortal man would not reach even if he exhausted his eternities counting, and whose imaginary dynasties have the letters of the Hebrew alphabet as ciphers. It was not given to me to enter that delicate labyrinth.[1]
>
> —*Jorge Luis Borges*, La cifra, *Madrid, 1981*

AT APPROXIMATELY the same time that Russian Orthodoxy was rent by the theological problem of Name Worshipping, the field of mathematics was also in turmoil, experiencing what the German mathematician Hermann Weyl would later call the "Grundlagenkrise der Mathematik" (the foundational crisis in mathematics).[2] These two stories, so different in their origins, came together almost a century ago in the discussions of Dmitri Egorov and Nikolai Luzin with their French colleagues. Dissimilarities between the French and Russian approaches would soon emerge.

The crisis in mathematics was brought about by the birth and rise of set theory in Germany in the last decades of the nineteenth century. A "set" is a collection of objects sharing some property and given a "name." For example, the set of all giraffes in South Carolina could be named "South Carolina Giraffes." This set obviously has a finite number of elements. By its description this set is different from the set of all flowers in your garden or all inhabitants of Cyprus, but in each case, the number of elements in these sets is finite. More interesting are sets with an infinite number of elements, such as the set of all integers (1, 2, 3, 4, 5, 6 . . .) later denoted **N**, where the ellipses mean that one thinks of the entire series of integers as potentially never ending. The set of all the points on a line segment is also infinite, but of a different sort. These examples raise the question of a definition of "infinity," something mathematicians had not managed to produce for over two thousand years. Yet, as Weyl observed, "mathematics is the science of infinity." Set theory tries to provide a framework in which all of mathematics can be fitted, and a definition of infinity was a crucial part of its elaboration.

Most non-mathematicians think they have some idea of what "infinity" means; they would probably say "something without end or limit." For mathematicians, however, the problem of infinity has been deeply puzzling. Discussions about infinity will play an important role in our story about French and Russian mathematicians in the early twentieth century.

How does one define infinity? Does it really exist, or is it only an abstraction? Is there only one "infinity," or are there several, perhaps many? Can some infinities be "larger" than others? Georg Cantor, a German mathematician, created set theory from his deep inquiry into these questions. Cantor gave infinity a mathematical definition after 2500 years of unsuccessful efforts, and the ultimate result of his labors was to make set theory the *lingua franca* of mathematics. The evolution of cultural conceptions of infinity before and after Cantor reveals the significance of his achievement.

The first glimmer of a conception of infinity probably came at the birth of civilization. Is it possible to fathom the first non-trivial thoughts of our ancestors millennia ago, watching the unbounded horizon, feeling time passing continuously from the past to an unknown and frightening future? When did they begin to conceive the

A Crisis in Mathematics

idea of unlimited space and time? And did they, from the start, combine this idea with a concept of the unlimited power of a divine or non-human being they thought was above them? Divine perfection eventually became synonymous with almightiness, that is, infinite might. Was infinity a divine prerogative from the beginning? In all probability we will never know. But we have some early clues in an ancient Greek word that combines all of these concepts: ἄπειρον, or *apeiron*.

This word is found in the first philosophical text of the Greek tradition, attributed to Anaximander of Miletus, who probably lived from 610 to 540 B.C.E. He described the ultimate material principle as *apeiron*, "the Infinite" or indeterminate; "something without bound, form, or quality." So from the start the word contains a contradiction: it attempts to express what is not expressible (the ineffable).

Hermann Weyl, one of the leading mathematicians of the first part of the twentieth century, was inspired by the history of the school of Pythagoras (c. 569–500 B.C.E.) and its fascination with infinity. He wrote:

> Aside from the fact that mathematics is the necessary instrument of natural science, purely mathematical inquiry in itself, according to the conviction of many great thinkers, by its special character, its certainty and stringency, lifts the human mind into closer proximity with the divine than is attainable through any other medium. Mathematics is the science of the infinite, its goal the symbolic comprehension of the infinite with human, that is finite, means. It is the great achievement of the Greeks to have made the contrast between the finite and the infinite fruitful for the cognition of reality. Coming from the Orient, the religious intuition of the infinite, the *apeiron*, takes hold of the Greek soul. This tension between the finite and the infinite and its conciliation now becomes the driving motive of Greek investigation.[3]

Weyl's phrase "coming from the Orient" was undoubtedly a reference to the many years that Pythagoras was thought to have spent in Egypt, studying the mystical arithmetic and geometric teachings of the priests of Memphis.

The Greek word *apeiron* contained three main ideas that persisted in later centuries:

- the limitlessness of space and time
- a non-rational, religious, or mystic aspect to infinity
- the indefinability and impossibility of description (ineffability) of infinity

All three of these characteristics are negative, defining what infinity is not (not limited, not rational, not definable) rather than what it is.

Aristotle (384–322 B.C.E.) introduced a distinction that was also usually accepted in later times: infinity is a potentiality, not an actuality. He noted that if one takes a line segment (one-dimensional space) it is possible to cut that segment in half, and then cut the resulting half in half again, and so on endlessly. As he observed, "It is always possible to think of a larger number: for the number of times a magnitude can be bisected is infinite. Hence the infinite is potential, never actual; the number of parts that can be taken away surpasses any assigned number." Aristotle's approach remained the dominant one for centuries. It lies at the heart of calculus and most other mathematical treatments of infinity until the time of Cantor. Even today, the idea of infinity as only a potential is the intuitive concept of the layperson, who knows very well that any specific number he or she mentions can always be exceeded.

Discussions of infinity often end up with paradoxes or antinomies. Aristotle summed up one of Zeno's several paradoxes in his discussion of the attempt of a fast runner to catch up with a slow one who has a starting lead:[4]

> In a race, the quickest runner can never overtake the slowest, since the pursuer must first reach the point whence the pursued started, so that the slower must always hold a lead.

In the pursuit, the increment between the pursued (often referred to as "Tortoise") and the pursuer (often dubbed "Achilles") grows smaller and smaller; but if the chase is described as a succession of moments when Achilles catches up to a place where the Tortoise was a short time earlier, the increment never entirely disappears. Thus, while everyone can conceive of a fast runner overtaking a slow one, he cannot do this, according to Aristotle's writing of Zeno's argument; thus we have a paradox.

These paradoxes are not so easy to answer or refute. Bertrand Russell said they were immeasurably subtle and profound,[5] and there is still debate about them today. Take, for example, the unusual answer proposed by Alexander Yessenin-Volpin (Aleksandr Esenin-Volpin), a Russian logician of the ultra-finitist school who was imprisoned in a mental institution in Soviet Russia. Yessenin-Volpin was once asked how far one can take the geometric series of powers of 2, say (2^1, 2^2, 2^3, ..., 2^{100}). He replied that the question "should be made more specific." He was then asked if he considered 2^1 to be "real," and he immediately answered yes. He was then asked if 2^2 was "real." Again he replied yes, but with a barely perceptible delay. Then he was asked about 2^3, and yes, but with more delay. These questions continued until it became clear how Yessenin-Volpin was going to handle them. He would always answer yes, but he would take 2^{100} times as long to answer yes to 2^{100} than he would to answering to 2^1. Yessenin-Volpin had developed his own way of handling a paradox of infinity.[6]

In part because of these paradoxes, the ancient Greeks had a fear of infinity; in fact, sometimes the word *apeiron* took on negative connotations such as "formless chaos"—something that should be avoided. Aristotle's description of infinity as only a "potentiality" rather than an "actuality" was one way to deal with this problem. For Aristotle, infinity could be conceived but never confronted directly.

Plotinus, a mystic and philosopher of the Neoplatonic school near the end of the classical period (he lived from 204 to 270 C.E.) faced infinity in a more positive way. He saw a correspondence between his god, the *One*, and an ineffable infinite. He argued that if the One was not infinite, then there had to be something beyond it, which was untenable to him. Interestingly, Plotinus's view linking the infinite and the divine in an affirmative way attracted the attention of one of the main figures in this book, the Russian mathematician Nikolai Luzin. In 1909 Luzin had recently recovered from an intellectual and spiritual crisis, brought on by both political and intellectual events. The Russian Revolution of 1905 had destroyed his earlier radical, materialistic ideology, and the work of Georg Cantor and that of French mathematicians (Lebesgue, Borel, Baire) had opened up new puzzles in mathematics. Tutored by Father Pavel Florensky, Luzin underwent a religious conversion, and subsequently, seeking further philosophical and religious assistance, he turned to Plotinus.

Because infinity was regarded as furnishing some knowledge of God, there was a fascination with Infinity during the medieval period. For example, Gregory of Rimini (1300–1358) anticipated that something that was infinite could be equal to a subpart of the whole infinite. Galileo saw something similar when he observed that there are as many integers as there are even numbers. If we write "1, 2, 3, 4, . . ." obviously we can continue indefinitely, and the entirety of such a series is infinite. But if we write only the even numbers "2, 4, 6, 8, . . ." just as obviously that series can also be continued forever and is infinite. Common sense would seem to indicate, however, that there are only half as many even numbers as there are all numbers (odd and even), so how could there be an equal number in both series? (In modern terms, doubling produces a one-to-one correspondence between the two sets.) Galileo concluded from this example that there could not be any definition of infinity.

A fuller and more positive approach was presented by Nicholas of Cusa (1401–1464), who advanced many prescient ideas, including a heliocentric view of the relationship of the earth and the sun. He also added metaphysical considerations to geometrical analogies to illustrate the meaning of infinity. Nicholas described circles of larger and larger diameters, showing that a segment of such circles approaches closer and closer to a straight line, as shown in this diagram:

A Crisis in Mathematics

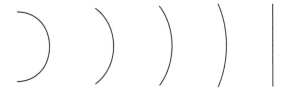

Larger and larger circles with segment approaching straight line, as suggested by Nicholas of Cusa.

A segment of a circle of infinite diameter would therefore coincide with a straight line. Nicholas maintained, however, that the rational mind cannot comprehend an infinite like this. Such understanding, he thought, could only be achieved through a mystical, religious insight.

An important further breakthrough was achieved by Bernard Bolzano (1781–1848), a Czech priest who made an in-depth study of the speculative tradition of medieval philosophy. Bolzano was alarmed by the "antinomies" of infinity. In his book *Paradoxien des Unendlichen (The Paradoxes of the Infinite)*, he tried to deal with various mathematical infinities in the same way as ordinary finite numbers. He introduced the word "set" *(die Menge)* and defended the concept of the "actual infinite." Furthermore, he stated explicitly that two infinities are the same if there is a law that assigns in a one-to-one way an element of the second set to each element of the first.

Set theory proper was created by Georg Cantor, who was born in 1845 in St. Petersburg, Russia, where his father was a merchant. The father, Georg Waldemar Cantor, had Jewish parents but was baptized in the Lutheran faith. In 1856, when the son was eleven, the family moved to Germany, where Georg the younger studied in schools in Wiesbaden and Darmstadt. Early on he demonstrated exceptional skill in mathematics, particularly trigonometry. He also studied and enjoyed philosophy and theology, and religion remained an important influence throughout his life. In 1862 Cantor entered the Polytechnic of Zürich, and soon convinced his father, who wanted him to be an engineer, that he should go instead into mathematics as a profession. He also studied at the University of Berlin and at Göttingen. After completing graduate studies in mathematics, in 1869 he took a position at the University of Halle.

Set theory began in December 1873 when Cantor proved, to his amazement, that the set of integers **N** (starting in 1895 Cantor would call its "number of elements" $\aleph_0$, aleph-zero) and the set of real numbers **R** (the Continuum) had different kinds of infinite numbers of elements. Thus it was natural to ask if there could be any set with a different infinity squeezed between the infinity of integers and the continuum of the real numbers. This is a weaker form of the Continuum Hypothesis (CH),[7] stated as early as 1879, which would be a source of work for Cantor for years and of mental troubles for some mathematicians, starting with Cantor in 1884 and including Baire and Alexandrov.

During a period of thirty years of intense work, Cantor built the foundations of set theory. In the first part of his research (1873–1882) he kept close to the original problem of subsets of the real line, but he obtained striking results, like the one already mentioned.

Cantor gave the first definition of a set only in 1883, when he wrote to Richard Dedekind:

> By a manifold or a set I understand in general every Many that can be thought of as a One, i.e., every collection of determinate elements which can be bound up into a whole through a law, and with this I believe to define something that is akin to the Platonic ειδος or ιδέα.[8] [Cantor favored, in Greek, these terms over Aristotle's only potential *apeiron;* he also quoted Plato's *Philebos.*]

Then Cantor started from the operation which associates to a part P of the line the set P', called the derived set, made up of its limit points. Cantor iterated the operation: for any part P, take P', then the derived set of P', call it P" . . . , and he went on iterating an infinite number of times and even further, "transfinitely." Cantor looked at this process and stated, "We observe a dialectical generation of

A Crisis in Mathematics

concepts which leads always further and which in doing so remains free from any arbitrariness, necessary and consistent in itself."[9]

Cantor collected a large number of examples like a naturalist picking up strange flowers, and in doing so, he discovered what was later called the "Cantor ternary set" which played a crucial role from then on (even in more recent years as an example of fractals). Cantor divided a line segment, say (0, 1), into three parts. Then he eliminated the middle third. Then, repeating this process, he eliminated the middle thirds from the two remaining segments, and continued on and on, as shown here:

Cantor ternary set.

The intersection of all remaining subsets constructed at each step is the ternary Cantor set. It has the same "number of elements" as the Continuum.

The first deep result concerning the Continuum Hypothesis came in 1884 when Cantor proved that any closed set either was denumerable or contained a one-to-one image of the Cantor set, thus proving that all closed subsets of the line satisfy CH. To Cantor, this was a clue that CH might be true in general.

Cantor also defined an infinite hierarchy of cardinals and ordinals, which can best be explained by an example. Let us imagine that a child has been given some apples. She may look at the apples and say, "One, two, three, four, five, six, seven. I've got seven apples." The first "seven" was uttered as the last apple was observed, but the second "seven" was intended as a statement covering the total set of apples. The first "seven" can be called an ordinal number (*Zahl* in Cantor's German), and the second "seven" a cardinal number *(Anzahl)*. Cantor extended the two notions to any set S (finite or infinite), assuming the elements are given in a certain order.

As an example, let ω (omega) be the ordinal number of the usual set of integers in normal order

1, 2, 3, ...

Put a different order on the set of integers by writing first even numbers, then odd numbers, and then add 1 at the end:

2, 4, 6, . . . ; 3, 5, . . . ; 1

Then this ordered set has ordinal number ω + ω + 1, but its cardinal is aleph-zero like all denumerable sets. Thus, if you give a different order to a set, you change its ordinal number but not its cardinal.

In Cantor's fundamental work of 1883, "Grundlagen einer allgemeinen Mannichfaltigkeitslehre" ("Foundations of a General Set Theory"), he developed metaphysical ideas on "Free Mathematics" in response to Leopold Kronecker's criticism. These ideas remained with him all his life and were related to his strong religious beliefs. Reflecting the influence of Spinoza, Cantor thought that mathematical concepts have an "immanent reality," based on well-definedness and non-contradiction, and a "transsubjective or transient" reality depending on representation in the external world. Both kinds of existence correspond to each other.

The new freedom for mathematics provided room for its future development and strongly impressed the Russians influenced by the Name Worshipping movement; it also stimulated the axiomatic school elaborated by David Hilbert in the 1930s. The Russians looked back at what Cantor had done and attributed a new significance to "naming" in mathematics; after all, Cantor had named a whole hierarchy of the alephs which possessed in his view an "immanent" existence. The Russians believed that the new infinities took on a reality after being named that they did not possess earlier.

Cantor was conscious of the importance of the steps he had taken. He wrote to Dedekind in the same letter referred to above:

> It has pleased Almighty God that I have attained the most remarkable . . . results in set theory . . . that I have found what

A Crisis in Mathematics

fermented in me for years and what I have long been searching for.

The reactions to the new theory were quite mixed in Germany: the younger generation of mathematicians was somewhat receptive, but the Berlin School under the leadership of Leopold Kronecker was strongly hostile. Kronecker believed that one day "all analysis and algebra will be founded on the strict concept of integer," and that all mathematics should be constructed in a finite number of steps. This opposition had a negative effect on Cantor's academic career and on his state of mind as well.

Charles Hermite, an important French mathematician, uncle of Paul Appell and father-in-law of Émile Picard, was not as rigid as Kronecker. He had been educated in traditional French philosophy and disliked discontinuous functions (functions that are not smooth, but have jumps or breaks), but he nonetheless admitted their existence. Hermite analyzed his own philosophical background in a letter of December 24, 1880, to the Swedish mathematician Gösta Mittag-Leffler:

> Analysis is for me largely a science of observation. Analysts seem to me to be naturalists who with the eyes of the mind look on a world as real as that of nature, at beings outside of themselves, which they have by no means created, and whose existence is as much in the necessity of things as animals and vegetables. The study of the subjective world allows therefore an insight, a view on the real world.[10]

Hermite admitted that the discovery of highly discontinuous functions could diminish the belief of "natural philosophers" in the complete continuity of the laws of Nature and cause them to modify their conception of the real world.

Among the younger German mathematicians, Paul du Bois-Reymond (1831–1889) also objected to a part of the new set theory. He accepted the "actual infinite" but rejected the philosophy underlying the treatment of the Continuum because Cantor was not making any distinction among sets; in contrast, du Bois-Reymond wanted to give the Continuum a kind of mystical status outside of mathematics. He believed that understanding the Continuum was beyond the capabilities of mathematicians.

Mittag-Leffler saw immediately the possible uses of Cantor's ideas in the theory of (analytic) functions, much more amenable to mathematical study than the problems of the Continuum. Mittag-Leffler became an active supporter of set theory, helping to spread Cantor's ideas among the important French and German mathematicians who were his colleagues and friends. One of these was Charles Hermite; another was Henri Poincaré, soon to become the star of French mathematics and later the patriarch often called "the last universal mathematician."

In 1882 Mittag-Leffler suggested that Cantor's work should be translated into French, but he was more positive toward Cantor's "mathematics" than his "philosophy." Hermite and Poincaré agreed that French readers of Cantor might object to "research which is at the same time philosophical and mathematical and where arbitrariness has an excessive place." Hermite thought that Cantor's work was more German metaphysics than mathematics. Perhaps for that reason he suggested as translator a Jesuit priest at Saint-Sulpice, observing that "[Cantor's] philosophical turn of mind will not be an obstacle for a translator who knows Kant."[11] During the translation work, Hermite and Poincaré made suggestions for cuts in the philosophical parts. And even some of the mathematics seemed too abstract to them; they observed, as Poincaré remarked, that "higher infinities" seemed to "have a whiff of form without matter, which is repugnant to the French spirit."[12]

Here Poincaré revealed his own philosophical approach to mathematics and mathematical objects: they cannot be purely abstract notions but must refer to material objects—bodies, planets, populations. This Aristotelian approach was a part of an old French tradition (Laplace, Fourier) which influenced the French school of mathematical physics and the main part of the French mathematical

A Crisis in Mathematics

school until the Second World War. This French approach would result in a disagreement between Poincaré and Hilbert at the International Congress of Mathematicians in Paris in 1900. Even the French mathematician Émile Borel, who used notions from Cantor's set theory for point sets, displayed an increasing mistrust about the generality of the theory. Despite this French suspicion of set theory, however, its concepts became widely known among mathematicians in the very last years of the nineteenth century.

In August 1897 another French mathematician, Jacques Hadamard, gave a talk in Zurich in which he suggested using set theory not just for the consideration of points in the Continuum but for the study of "sets of functions." This proposal was motivated by the problems of the calculus of variations and had been undertaken simultaneously by the Italian School. About ten years later this direction would be given further development by Maurice Fréchet, a student of Hadamard, leading to the birth of functional analysis.

The French mathematician René Baire took the opportunity of a fellowship to visit Turin and subsequently made a breakthrough in the modern analysis of functions. He constructed a classification of most continuous and discontinuous ones. Baire's 1899 thesis on this subject caused quite a scandal in the French mathematical establishment. One of his thesis committee members, Émile Picard, out of respect for Baire's obvious mathematical ability, managed to approve the thesis—along with the other members of the committee, Darboux and Appell—but he was skeptical about Baire's entire approach. Indeed, Picard was suspicious of all efforts to mix philosophy and mathematics. Baire was more receptive to such attempts, but even he had some doubts, as did many in the French School.

Among the French mathematicians, the leaders in wrestling with the implications of set theory were Émile Borel (1871–1956), Henri Lebesgue (1875–1941), and René Baire (1874–1932), referred to in this book as "the French Trio." They went on, following Cantor, to write an illustrious page in the history of mathematics, but they would later be assailed by doubts about what they were doing. Eventually they came to an intellectual abyss before which they halted.

{ 31 }

Faced with this frightening prospect, and influenced by the rationalistic culture in which they lived, they lost their nerve, each expressing this frustration in a different way that revealed much about their individual personalities.

But before that happened, they taught what they had learned about set theory to the Russian mathematicians Dmitri Egorov (1869–1930) and Nikolai Luzin (1883–1950), who, together with their friend Pavel Florensky (1882–1937), make up "the Russian Trio." The Russians, invigorated by the mystical belief in the power of Name Worshipping which the Mt. Athos monks had spread throughout Russia, managed to make their way across the abyss. In the different reactions of the French and the Russians to set theory, the impact of their distinct cultural and religious traditions became very evident.

3

The French Trio: Borel, Lebesgue, Baire

> "Who of us would not be glad to lift the veil behind which the future lies hidden, to cast a glance at the secrets of the development of mathematics during future centuries?"
> —David Hilbert, *International Congress of Mathematicians, Paris, August 8, 1900*

THE MOMENT at which the importance of set theory for mathematics became obvious, especially for the French, was the second International Congress of Mathematicians held in Paris in 1900. This gathering was kicked off with an address by the German mathematician David Hilbert (1862–1943), a speech which was immediately recognized for its significance and which today is seen as perhaps the most famous speech in the history of modern mathematics. Hilbert showed clearly that he thought Cantor's set theory would play an important role in the future of the field. He placed CH, the Continuum Hypothesis, at the top of his list of 23 major mathematical problems.

At this time Europe was in the midst of the "Belle Epoque," a time of stability, peace, and prosperity. Underneath the surface appearance of well-being there were, of course, problems: international tensions, especially between Germany and a France still lamenting the loss of Alsace-Lorraine in 1871, and a growing economic struggle between the affluent and those at the bottom of the economic order. The Second International was uniting socialist and labor parties, criticizing colonialism and imperialism, and envisioning an entirely different economic system. In only five years Russia would be racked by a violent, unsuccessful revolution, which would be a turning point in the life of the most important Russian mathematician in this book, Nikolai Luzin.

But in August 1900, Paris seemed to be the peaceful and beautiful center of the cultured and civilized world. Mathematicians arrived for the congress from all over Europe, and a small number of them took advantage of their trip to Paris to attend some lectures at the International Philosophy Congress at the Sorbonne, just before the mathematics meeting. There they heard Émile Boutroux and his brother-in-law Henri Poincaré talking about the philosophy of science. During their time in Paris the younger mathematicians strolled late at night along the river banks, where they admired the pavilions of the Universal Exhibition which had already attracted millions of tourists to Paris, and the recently built Eiffel Tower. Later, at the end of the congress, some of the participants would indulge in an evening at the Théâtre de la Renaissance to see Sarah Bernhardt, "the Divine Sarah," the most famous actress in the world.

David Hilbert, a professor in Göttingen, had accepted an invitation to deliver the opening address at the Paris Congress from his "friendly" competitor Henri Poincaré. At the first International Congress three years earlier, Poincaré had given a general talk emphasizing the connection of mathematics with the exact sciences; Hilbert, on the advice of his colleague Hermann Minkowski, a young genius teaching in Zurich, decided to take up the gauntlet and respond to Poincaré with a different view of mathematics. Hilbert did not address the relevance of mathematics to other fields; instead he discussed the problems mathematics faced within itself.

The competition here was not only between the world's last two universal mathematicians, but also between two philosophies of math-

The French Trio

ematics: Poincaré represented the old French ideology (that of Fourier, Laplace, and many others), viewing mathematics as closely connected with physics and the world; Hilbert propounded a different ideology, one that was closer to Kant and more abstract. In short, this was a rivalry between the two main schools of mathematics of the time, French and German, with an obvious nationalist dimension.

Hilbert gave full recognition to set theory and Cantor's work since the 1870s. As described in the previous chapter, set theory began with Cantor's proof that there are at least two different infinities: a denumerable infinity made up of the infinite number of integers (later named $\aleph_0$, aleph-zero, by Cantor) and the non-denumerable infinity of points on a line (the Continuum).[1] Cantor's original argument was based on Eudoxus' approach using nested sequences. The "diagonal argument" he used in 1878 showed not only that there are "more" points in the Continuum than in the set of integers but that there is a strictly increasing series of infinities starting from $\aleph_0$—a whole hierarchy of infinities, an infinity of different infinities.

Cantor had hoped to complete the classification of these infinities by showing that the Continuum (the set of points on a line) was the next aleph after $\aleph_0$—the Continuum Hypothesis. He never stopped working on this problem in the 1880s and 90s, but the increasing difficulties he faced were partly responsible for the series of mental crises he suffered after 1894. Hilbert in his 1900 Paris speech emphasized the importance of the problem of the Continuum, an issue that had obsessed and bedeviled Cantor.

Cantor's Reception in France

In 1900 three young mathematicians in France, Émile Borel, René Baire, and Henri Lebesgue, took full notice of Hilbert's talk at the Paris Congress. Lebesgue was at the time teaching mathematics in Nancy and was probably too poor to make the trip to Paris. Borel, however, lived in Paris and was sitting in the audience when Cantor gave his introductory speech. He had taken with him the younger mathematician René Baire, for whom he was developing a strong affection.

These three men would shape much of the French response to set theory, and in the process would make fundamental contributions to mathematics. The origins of their attitudes can be found in the details of their lives, as well as in the intellectual milieu of France—the land of René Descartes and Auguste Comte, and the home of a strong rationalist tradition. Every schoolchild in France was taught the words of the seventeenth-century literary critic Nicolas Boileau, "Anything that is understood well can be expressed clearly, and the words then come easily."[2] This did not leave much room for ineffability!

Descartes was a dominant influence among French thinkers and scientists. According to Descartes, thinkers should work to

> divide each of the difficulties under examination into as many parts as possible, and as might be necessary for its adequate solution. To conduct my thoughts in such order that, by commencing with objects the simplest and easiest to know, I might ascend little by little, and, as it were, step by step, to the knowledge of the more complex, assigning in thought a certain order even to those objects which in their own nature do not stand in a relation of antecedence and sequence.

Thus, every problem should be broken down into its simple components, and thought means clarity and expressibility.

One can see the strength of Descartes's influence in the numerous articles and speeches that honored his memory at the turn of the nineteenth and twentieth centuries. For example, Picard, an outspoken opponent of set theory, proclaimed on the three hundredth anniversary of Descartes's birth in 1896: "I always had, as is appropriate, an infinite respect for Descartes. One must judge Descartes on the completely new orientation he gave to science by his genius-like intuitions and by his method."[3]

Descartes was also an influential proponent of the view that mathematics was the most universal and least biased form of knowledge.

The French Trio

Most French mathematicians wanted, as far as possible, to segregate philosophical and mathematical questions. In contrast, the Russian mathematicians who later learned set theory from Borel and Lebesgue wanted to integrate philosophical—indeed, religious—issues with mathematics.

A second important influence among French mathematicians was positivism. The end of the nineteenth century saw the triumph of Auguste Comte's positivism not only at the Sorbonne but throughout the French educational system, which was reformed along Comtian lines in 1902.[4] According to Comte, once science liberates itself from all metaphysical influences and enters the "positive stage," its goal is no longer a metaphysical quest for truth or a rational theory purporting to represent reality. Instead, science is composed of laws (correlations of observable facts) that can be used by the scientist without regard to the nature of reality.

Émile Borel led an intense and very active life. He was so gifted that his wife Camille would write after his death that he was "consumed with all possible experiences, he threw himself into existence as a swimmer dives into water, he was devoted to science, to his friends, to politics, to the pursuit of joys of the most diverse kinds." At different times he was a brilliant young mathematician; professor; socialite in Paris; director of the École Normale Supérieure; journalist; publisher of the *Revue du Mois* (which played a crucial role in the formation of the Radical Left); friend of Edouard Herriot and Painlevé, and one of the leading figures of the Radical Party; mayor of his small home town; chief of scientific and technical services for the Ministry of War during the First World War; Minister of the Navy for six months; activist in the French Resistance; prisoner of the Gestapo; and holder of innumerable honors and decorations. For almost his entire life he was helped by an exceptional intellectual bond with his wife Marguerite, known as Camille Marbo (*Mar*-guerite *Bo*-rel), later a "femme de lettres," a woman he, at age thirty-one, carefully chose when she was eighteen.[5]

Borel was born in the small village of Saint-Paul de Fonts in the *département* of Aveyron in rural southwest France. Honoré Borel, Émile's father, had a property there in a landscape of small hills with a dry climate, typical of the south of Rouergue and quite similar to Corsica or other parts of the Mediterranean coast such as Greece or Algeria. The nearby larger village of St. Affrique was named after the seventh-century St. Africus (who reportedly was buried in the area). St. Affrique, located on the Sorgue River, is surrounded by beautiful bright green hills, red earth, and deep river gorges. The history of the area is as colorful as its geography, characterized by religious and military conflict.

The reputation of the area for dissent was strengthened during the religious wars of the seventeenth century, when St. Affrique was a stronghold of French Protestants, the Huguenots. Although they were defeated by a royal army in 1629, their influence remained. Émile Borel's father Honoré was, in fact, the Protestant minister of the town and the creator of a free Protestant school; the future mathematician lived in his parents' large house in the center of the commune next to his father's church, a short walk from the Pont Vieux, one of the most beautiful medieval bridges in France.

Borel loved his home region deeply and believed that it was rooted in wisdom and true human feelings going back many centuries. Long after he moved to Paris, and indeed well into old age, he would return frequently to St. Affrique, where he would be seen walking the hills and admiring the fields, the workers, and the cattle. He never forgot his homeland in the heart of Rouergue, combining this local "point of anchorage" with strong French patriotism and belief in the defense of the nation. Borel's young friend and student Arnaud Denjoy (1884–1974) remembered his teacher as having "numbers and earth from Rouergue glued to his shoes" ("à la semelle de ses souliers"), and perhaps for this reason Borel would later have difficulty accepting concepts in set theory that he thought could not be tied to anything "real."

Borel was a brilliant young pupil in his father's school and quickly revealed a strong appetite for both life and academic knowledge. In the France of the Third Republic some of the principles of the

The French Trio

French Revolution were still alive, including laicism, republicanism, and an emphasis on education (as Danton said, "Food first, then just after that, education").

The French government provided an upward ladder for bright, hard-working boys like Émile Borel. He seized the ladder with both hands and ascended as rapidly as he could. His brilliant results at the St. Affrique school gained him admission to the elementary mathematics class in one of the best of the Paris lycées, Louis-le-Grand, next to the Sorbonne and the Collège de France; he then advanced to the special mathematics class of M. Newenglozski, doctor of mathematics and later general inspector. In this group he became friends with the son of Gaston Darboux, one of the leading mathematicians of the previous generation. At the age of 18, Borel placed first in the entrance examinations for the École Normale Supérieure (ENS) and the École Polytechnique, both excellent institutions in Paris. Borel, influenced by Darboux's example, was already thinking about a possible academic career, and for that reason he chose the ENS, on the rue d'Ulm in the Latin Quarter, which he entered in 1889.

Borel continued to excel in his studies, graduating first in his class in 1893. He then taught at the University of Lille for a short time. In just a few years there he wrote not only his thesis but many articles that showed he was destined to be one of the leading mathematicians of his generation. He also served in the army for a year, teaching mathematics to young soldiers.

After military service Borel began preparing a dissertation at the Sorbonne with Gaston Darboux. His subject, the classical theory of functions and the distribution of their values, was in the mainstream of the French school in that period. Borel's research fell within the framework created by Augustin Cauchy (1789–1857), "the father of modern analysis," and Charles Hermite (1822–1901), who had further developed Cauchy's study of infinitesimals and the technique of computing with them. Cauchy founded the concepts of mathematical analysis (starting with notions of limit or continuity) on an arithmetical basis, a point of view that was pursued in France by Charles Hermite and in Germany by Karl Weierstrass (1815–1897) under the name of "arithmetization of analysis." The latter term was introduced by Felix Klein and Leopold Kronecker (1823–1891), who radically pursued this line by the use of explicit constructive methods. Kronecker said, "God created the integers; all the rest is the work of man"; he believed that only finite steps could be taken, excluding irrational numbers, which for him did not exist. Kronecker firmly opposed Cantor's set theory.

When Borel began his work, the arithmetization of analysis was accepted by most mathematicians. Hermite had also developed the ideology that the only functions that should be considered were "smooth" (continuous) ones; he despised "that lamentable plague of functions without derivatives."[6] Borel, however, at the time of his first research with Darboux had to consider the limits of points, and showed his creativity in using the theory of sets to do so. (Set theory had already been timidly introduced to France by Camille Jordan in a course at the École Polytechnique). Borel proved a key result concerning any covering of a fixed interval by an infinite sequence of small intervals ("Theorem of Heine-Borel"). In fact, this result created the basis of the future theory of "Borel measure," starting in a publication of 1898. Later Lebesgue would further develop this concept.[7]

Borel's interest in set theory began with what he at first called a "romantic attraction," although like many such attractions, it would later cool. Borel subsequently excused this early fascination with set theory by observing,

> Like many of the young mathematicians, I had been immediately captivated by the Cantorian theory; I don't regret it in the least, for that is one mental exercise that truly opens up the mind.

All French mathematicians at that time believed that the results from Cantor's theory on the limits of points would be useful only in the study of functions, that is, in the natural framework of the theory of analytic (very regular) functions.

The French Trio

When Borel submitted his brilliant thesis in 1894, the members of the jury were very much to his taste and character: Gaston Darboux, the father of his classmate; Henri Poincaré, the leading mathematician of the time; and Paul Appell, an important mathematician, close friend of Poincaré and later rector of the University of Paris (and also Borel's future father-in-law). Soon after the defense, Borel was invited to join the faculty of the ENS. Just a few years later, in 1898, Borel published his lectures with the results of his thesis on the theory of functions (there would be three more editions among his works of more than thirty books).[8] He presented there a detailed description of set theory and new notions of measure. Borel was attracted to "down-to-earth" problems such as the measure of the length of a circle: how do you define its length if you only know the length of a segment? Obviously, you need some means of "passing to the limit," of making approximations, as Eudoxus and Archimedes had done more than 2000 years earlier. Extending and deepening their point of view with the vocabulary of set theory, Borel defined a new class of domains, "measurable sets," later known as B-sets or "Borelian sets," to which a measure could be assigned.

Borel now joined not only the scientific life of Paris but also its vibrant social life, and soon ended his bachelorhood. The family of Paul Appell, including his wife and three children, were living in a small "hôtel particulier" on Rue Le Verrier, at the border of the Latin Quarter and Montparnasse. Many well-known visitors were coming regularly for dinner—Painlevé, Darboux, who also brought Clémenceau, and the uncle Joseph Bertrand, along with brilliant young mathematicians. This was where Borel met Marguerite Appell, daughter of his thesis examiner, who was thirteen years younger than he. Marguerite took immediate notice of this tall, handsome young man with brown hair and a beard. She later wrote that "he liked to dance and did not turn away from the pleasures of the world." Borel knew what he wanted, and later confessed to Marguerite: "I was watching you because you were different from the other girls." Émile saw that Marguerite was interested in deep questions, not just the gossip about fashions and society that attracted so many of her friends. On the other hand, Émile's obvious enjoyment of dancing, conversation, and the other attractions of Parisian life made his intellectual depth all the more appealing to the young girl and future feminist.

Émile's marriage to Marguerite in 1901 when she was 18 further strengthened his place in Parisian society and his integration into the most powerful family of mathematicians in France through the process of "natural families," as described by Raspail in 1837:

> After one intrigues for one's own benefit one does it for one's children, then for one's sons-in-law, then for their children still in the cradle; the system of natural families invades all the sanctuary, and a son-in-law, if directed by an omnipotent arm, must be quite clumsy if he gets beaten by a non-indigenous parvenu.[9]

The marriage gave further fuel to the ironic gossip among mathematicians that "genius is transmitted through sons-in law." After all, Marguerite's father, Paul Appell, had married a niece of the mathematician Joseph Bertrand, himself the brother-in-law of Hermite (and Picard in turn was a son-in-law of Hermite). Marguerite's mother had among her ancestors two good Jewish mathematicians from the beginning of the century. But Marguerite was a creative person herself, and she soon built a brilliant career as a feminist writer of the Belle Epoque.

After their marriage the Borels met and became friends with a group of the Parisian intelligentsia, first meeting once a week in their small flat on the fifth floor of 30 Boulevard St. Germain. Their close friends included Paul Montel, Henri Lebesgue (who came from Rennes), Paul Langevin and Jean Perrin (both famous physicists and popularizers of science), Paul Painlevé (future prime minister), Émile and Pierre Boutroux (father and son, mathematician and philosopher), Jules Tannery (mathematician and brother of the historian of science Paul Tannery), and Charles Seignobos (a famous historian

The French Trio

who built a vacation place at L'Arcouest in Brittany, later acquired by his friend Borel). After the tragic accidental death of Pierre Curie in 1906, his widow Marie became a very close friend of Camille Marbo-Borel.

Often Marguerite Borel would hold a "soirée" in their subsequent bigger flat or, even later, in their enormous apartment at the École Normale, where they settled after 1901. There, in an abode where Louis Pasteur used to live, Camille would be the mistress of the evening, entertaining the stars of the Parisian cultural and intellectual firmament. Thus Émile Borel, coming from a remote rural part of France, through a combination of achievement and good luck rose very high in just a few years. But despite his emergence as a social and intellectual figure in Paris, Borel kept his connections to his distant hometown of St. Affrique, first by becoming the owner of his father's farm, with a hundred and fifty sheep, eight beef cattle, honey bees, and other farm animals. He made a good profit there until World War I, and often received his Parisian friends at his country place.

Borel began teaching at the ENS and was scientific director from 1901 to 1918. Camille very much enjoyed the life there; their apartment was so large that Borel could have lunch with eight *normaliens* and two teachers while Camille would be quietly reading or gossiping in another part of their home. When Marie Curie as a widow in 1911 was the victim of a public scandal as a result of her affair with the married Langevin, she took refuge in the flat with the Borels, and had two rooms for herself.

Later, after 1914, Borel stopped doing mathematics and devoted himself to administrative duties, becoming in 1924 a deputy of the political left (the "cartel des gauches"). He even briefly held the post of Minister of the Navy in 1925 (an important fact for his later Soviet critics), but he always kept local elected positions as well. Borel enjoyed the duality of provincial and Parisian life, and once said, "What I like is having lunch on Sunday with the old Rigaud in Broquiès [the mayor of a small village near his rural home] and the

Émile Borel.

next day having a discussion with President Doumergue at the Elysée Palace."[10]

In the early classes that Borel taught at the ENS, he encountered two exceptional students, Henri Lebesgue and René Baire. The two came from poor families but had risen through educational achievement; Borel, while quite well off, had also pulled himself up by his talent. Moreover, the three men were united by a common love of mathematics and shared other tastes and interests as well. Thus the "French trio" was formed.

Henri Lebesgue was born in 1875 near Beauvais, where his father was a typographer who provided the family with only modest resources. Three years later the father died, and Henri's mother was forced to take in sewing in order to maintain the household.[11] The fact that Lebesgue became one of the great mathematicians of France shows, as was also true of Baire and Borel, that the educational system of the Third Republic was effective in identifying and promoting talented young boys throughout the country.

Beauvais, where Lebesgue was raised, is a small city about sixty miles north of Paris, located near wooded hills on the left bank of the river Thérain where it meets the Avelon. It is the capital of the Oise *département*, with a beautiful cathedral built in the thirteenth century. In the seventeenth century another well-known mathematician, Gilles de Roberval (1602–1675), was born in the same area. Roberval was a quarrelsome and unbending character about whom Lebesgue wrote, perhaps thinking of himself, "He was described as envious and conceited . . . but without doubt he was a man of heart, combating all injustice inflicted on others as well as himself, but who in the fire of combat was carried away by passion and overshot his goal . . . This kind of disposition is not rare in the Oise region."[12]

Lebesgue was a very good pupil at the local school, but could not afford to prolong his studies. The mayor of Beauvais, E. Gérard, had known Lebesgue's father, an active socialist who had created many cultural associations. Gérard helped Lebesgue to enter college, where he soon demonstrated both a passion and a great talent for mathematics, especially geometry; in fact he was later called the "aristocrat of geometry" because of the elegance and purity of the ideas he presented. He preferred to see all mathematics in geometric terms. Excelling in his examinations, Lebesgue was admitted to the École Normale Supérieure in Paris, where he met Baire and Borel.

The ENS became a center of protest during the Dreyfus Case (L'Affaire Dreyfus), the famous controversy that raged in France in the years 1880–1905 around the charge of treason levied against the Jewish army officer Alfred Dreyfus. The affair became known throughout all of Europe and even had an impact on the Russian mathematician Dmitri Egorov, who lived in Paris at the height of the controversy. Anti-Semitism existed in both France and Russia, as well as in many other countries at the time.

The French mathematics community became deeply involved in the Dreyfus Affair when Paul Appell was convinced by Jacques

Henri Poincaré.

Hadamard (a distant cousin of Dreyfus) that the army officer was innocent. Another mathematician (and later famous politician), Paul Painlevé, drew up in 1898 a moderate petition ("appel à l'union") decrying the accusations against Dreyfus, but not all of his fellow mathematicians would sign it. Lebesgue, a kind of intellectual anarchist who distrusted the power establishment—in this case, the army—readily signed, but Borel, a patriot, refused, still trusting the military authorities. Borel, it turned out, was capable of changing his mind on the matter, but some other mathematicians like Hermite and Picard (member of a rightist league) were firmly among the "anti-Dreyfusards."

The leading French mathematician of the time, Henri Poincaré, was not very interested in politics (unlike his brother Raymond Poincaré, the future prime minister), but he gradually got drawn into the debates and eventually played a leading role. After the first trial of

The French Trio

Dreyfus and his sentencing, more and more intellectuals began to believe that a great injustice had been done as a result of pressure from the army. Borel managed to overcome his patriotic loyalty and joined the critics. Appell, Darboux, and Poincaré formed an independent "jury" for the purpose of studying the evidence against Dreyfus, which rested almost entirely on a handwritten note ascribed to the army officer on the basis of pseudo-probabilistic arguments. Poincaré, master of probability theory, thoroughly destroyed the arguments advanced by the army authorities. Thus the affair involved mathematicians in a deep way and left a lasting influence.

Like Borel, Lebesgue became interested in functions, but differed in his devotion to geometry. He later commented, "There are connections which I feel to be very close between the general theory of functions of a real variable and pure geometry, but they remain a little mysterious to me." As a student at the ENS, he was whimsical and carried on the tradition of hoaxes. Even years later Borel's wife, Camille Marbo, would remember Lebesgue's "malicious smile under a reddish moustache." His fellow students noticed that he often carried around with him a sheet of paper which he would fold and crumple in various ways to show the properties of surfaces. Lebesgue would explain that the properties of *ruled* surfaces, like a sheet of paper, cannot remain true in all generality. He would exclaim, "You see, with wrinkling the *rule* disappears!" It was exactly this kind of geometric spirit applied to analysis which in 1901, at the age of 26, led him to the achievement for which he is probably best known, the "Lebesgue Integral." The construction of the integral starts with a simple geometric trick (and then proceeds using previous remarkable work by Borel).

At the ENS Lebesgue was strongly influenced by his teacher Borel, who taught him, right from the start, set theory and measure theory. This last topic would foster a strong competition between the two men, which started on friendly terms but would become a pretext for their later quarrels. At the beginning, however, Borel was a real role model for Lebesgue, who admired his diplomatic charms, his savoir faire, and his sociability. But it was clear by the time of the Dreyfus

Henri Lebesgue.

Affair that their views of the world differed. Furthermore, Borel was becoming frustrated by the difficulties of set theory. Lebesgue, however, bravely picked up the baton, not only from Borel but also from his fellow student René Baire (with whom, unfortunately, one of Lebesgue's numerous quarrels started very early).

Lebesgue, who was rapidly becoming a master of elegant and profound mathematics, started from the root question: what *is* a function? Functions are expressed as formulas; but sometimes functions have no explicit formulations, or not even implicit ones. Reflecting about this in conjunction with the classification of functions that Baire had given in his thesis in 1899, Lebesgue started with the nicest continuous functions (Baire called them "of class zero"), then obtained functions of class one and pursued this process "transfinitely," using new Cantorian notions, to increase the realm of functions considered. Thus Lebesgue started to construct more precisely than Baire had the classification of functions and sets. The result was a remarkable article of 1905 entitled "Sur les fonctions représentables analytiquement."[13] In this article the concept of "named" *(nommé)* mathematical objects emerged clearly.

Lebesgue had in fact used the word *nommé* earlier than 1905; it appeared for the first time in his "Leçons" of 1904. He remarked, "I do not know if it is possible to name even one function that is not B-measurable; I do not know if non-measurable functions exist."[14] Here Lebesgue was going beyond his master Borel, who did not use the expression "name," and even tried to avoid using transfinite numbers. Borel showed his guarded admiration and even skepticism about Cantor in this statement:

> Getting rid of them [the transfinite numbers] one gains in simplicity and clarity. This remark does not diminish at all the philosophical interest or the real importance of the profound ideas of M. Georg Cantor, whose influence on the evolution of mathematics in the last quarter of the 19th century has been, as we know, enormous; this influence will remain as long as there remain mathematical analysts, even if some particular forms in Cantor's thought might one day keep only a historical interest.[15]

Lebesgue was much bolder in his article of 1905. Not only did he emphasize the importance of naming, but he showed through a difficult proof (used eleven years later by Alexandrov and Hausdorff) that there exists a function in each class of the classification of Baire. Lebesgue tried to be as precise as possible, and he complimented Borel, saying:

> I will try never to speak of a function without defining it effectively; I take in this way a very similar point of view to Borel. . . . An object is defined or given when one has said a finite number of words applying to this object and only to this one;

that is when one has *named* [*nommé*] a characteristic property of the object.[16]

These remarks by Lebesgue are in complete harmony with his later position concerning the Axiom of Choice, discussed later in this chapter. They also reflect the philosophical debates in Paris in these years, for example, the discussions about Bertrand Russell in the *Revue de métaphysique et de morale* and the emergence of new paradoxes in set theory, as well as the debate between Couturat and Poincaré. These issues would make a strong impression on the young Russian mathematician Nikolai Luzin, who arrived in Paris in December 1905.

On the one hand, Lebesgue developed and gave a systematic extension of Borel measure with his "Lebesgue Integral," which had an immediate success worldwide; on the other hand, he pushed the classification of functions elaborated by Baire, thus moving into opposition to the old school (represented by Camille Jordan, Gaston Darboux, Charles Hermite). In his work of 1904 Lebesgue bravely grappled with all the "monsters" of discontinuous functions feared by Hermite and others, and thereby took what is really the first step toward descriptive set theory. Lebesgue went to the limits of what could be said at the time about the most general functions.

In almost all of Lebesgue's work there is a common thread made up of geometric intuitions. But in fact many aspects of set theory could be considered from a geometric point of view, such as the famous problems of the Continuum, or questions of measure. Moreover, Baire had introduced a space of infinite dimension, the space of all infinite sequences of integers, called later Baire space, and he showed that one could identify this space with the set of all irrational numbers. This appeared to be a new connection between set theory and geometry. So there were many opportunities for Lebesgue's geometric talents to develop.

René-Louis Baire was born in Paris in 1874. In a sense, however, his origins were even further from the centers of French intellectual life than Émile Borel's had been in remote St. Affrique in the rural south.

The French Trio

Baire's father was a tailor, and René lived with three brothers in financial circumstances considerably worse than those of Borel. Like Borel, Baire demonstrated brilliance early, reading Musset, Lamartine, and Chateaubriand, learning to play the violin, and listening to concerts in the garden of the Palais-Royal—but from afar, because he could not afford the ticket price of one franc.[17] In addition to his material circumstances, he had problems with his personality and with his health. As early as age 14, he began to suffer from digestive problems that affected him all his life. An acquaintance from his youth described him as "a big fellow with an obviously weak bone structure, a wan complexion, and dark deep eyes that tended to stare in a disturbing manner."

In 1886, at the age of 12, Baire won a scholarship that changed his life, since his family would not have been able to afford a good education for him. The scholarship gave him entrance as a boarding student to the Lycée Lakanal, an excellent school located in the Parisian suburb of Sceaux. Here Baire found a rich pedagogical environment in which he prospered. He stopped playing the violin and "replaced it with equations," his brother said. He twice won honorable mentions in national competitions with top students from all over the country. This performance won him further access to advanced mathematics classes in the Lycée Henri IV, and then admission to both the École Polytechnique and the École Normale Supérieure. He chose the ENS and found himself in lectures taught by Borel, Charles Hermite, and Émile Picard, as well as, at the nearby Sorbonne, those of Henri Poincaré—in other words, the mathematical elite of France.

In 1898 Baire managed to win another scholarship that allowed him to study in Italy, at the invitation of the leading Italian mathematician, Vito Volterra. Volterra, along with several other Italian mathematicians who had read Cantor in German, in particular Giuseppe Peano and Ulisse Dini, was working in mathematical analysis and was exchanging ideas with the French mathematician Jacques Hadamard.

Baire's personality problems continued to plague him. On ad-

René Baire.

vanced examinations, such as that for his "agrégation," he ranked first in the written parts but did less well on the oral examinations. His examiners seemed merciless, at least to him. Baire quickly developed a sense that the world was not fair to him. His first appointment, to a lycée in Bar-le-Duc, was not as distinguished as his ability in mathematics warranted. Bar-le-Duc was a small town situated in the Lorraine, far from Paris. Baire's teaching load there was heavy, but he somehow found time to continue his research in mathematics.

Baire's life was characterized by rigor—both in mathematics and in his life-style. He had a strict sense of duty and an immense respect for science. This rigor led him to think in a new way about the notion of function in mathematics.

Earlier mathematicians had various views on functions. In mathematics a central notion of functions emerged slowly, first through the algebraic considerations of Descartes, then in a more general setting but with strict limitations in the hands of Leonhard Euler (1707–1783). For Euler a function had an explicit expression; in particular, he believed that functions must be continuous and "smooth." Lejeune-Dirichlet was the first to consider arbitrary general functions with no explicit description, and Darboux began early in 1875 to study non-continuous functions. Baire advanced a new view. At his *agrégation* examination in 1895 he realized that there was no obvious answer to a problem involving functions of two variables, and this led him to the new notion of "semi-continuity" (assuming continuity, but only from the left or from the right) and then to a very original step forward: he succeeded in characterizing discontinuous functions that are limits of continuous ones (a little later they would be called functions of Baire class equal to one). His thesis along these lines was a masterpiece, and was the first step leading to future descriptive set theory. Denjoy later described Baire's work in this way: "In order to guess the precise statement one needed real gifts for observation, but to prove it you needed to use in a new context the Cantorian transfinite numbers." (Denjoy, the son of a wine merchant from Perpignan, would maintain close relations through the years with Borel and Baire and would also become Luzin's closest friend in France.)

With better health and better opportunities, Baire might have made much more progress in set theory than he was able actually to do. But shortly after his thesis defense his mental and physical difficulties reached the point where, for long periods, he was no longer able to work (a physician wrote of neurasthenic troubles in a 1900 certificate). As a result, he was often in financial difficulties. Just what the nature of his maladies was remains unclear. He had problems with his esophagus, and he developed a psychological disorder that, by his own description, "debilitated" him. The paradoxes and inherent complexities of set theory may have accentuated his health problems. His family and friends believed that his feelings of frustration about not being sufficiently recognized for his achievements played an important role. Increasingly bitter, he fell into a deep depression.

Arnaud Denjoy.

Baire's work on the theory of discontinuous functions of a real variable was an impetus to Lebesgue, who was able to define his integral for all the bounded discontinuous functions introduced by Baire. Lebesgue gave credit to Baire for his contributions, but Baire always considered it odd that Lebesgue in later years received positions at the Sorbonne and the Collège de France while he did not. And in pushing farther than his teacher, Lebesgue also created tensions with Borel: later in a letter to Borel he even criticized the latter's "career as a son-in-law."

Clouds in a Blue Sky, 1900–1904

Contradictions are called paradoxes or antinomies in philosophy, and they appeared very early in human thought—for example, the famous paradoxes of Zeno as described by Aristotle. Similarly, contradictions arose within the framework of Georg Cantor's mathematical theory of infinity. In the first years of the twentieth century Cantorian set theory was beset by a series of plaguing paradoxes which even now can cause headaches. Some of these difficulties were apparent as early as the 1880s, at least to Cantor, but he kept them to himself. His preoccupation with them may have been one more reason for his increasing mental problems.

Already in 1895 Cantor realized that there were difficulties with what he called "sets that were too big to correspond to any cardinal" (he took as an example "the totality of everything conceivable"), and he escaped from the resulting contradiction by introducing pluralities too big to be sets, corresponding to a theological notion, the "Absolute," which cannot be known, not even approximately. Other mathematicians exploring Cantorian set theory were not satisfied with such a theological solution to the difficulties, which were soon called "antinomies" in reference to Kant's *Critique of Pure Reason*, where Kant pronounced that there are inevitable contradictions when man is confronted with all-embracing notions like causality, freedom, or God.

In 1897 Cesare Burali-Forti showed that the concept of a set of all ordinals leads to a contradiction, thus essentially stating more clearly what Cantor had already realized. But the real blow came in 1901 (published in 1903) when Bertrand Russell, in what is now known as Russell's Paradox, analyzed the concept of the "set of all sets which do not belong to themselves" and explained the contradiction in simple words, so that it became quite popular. Russell's Paradox was very similar to the one constructed about the truth-value, in a logical sense, of a sentence attributed to the Cretan Epimenides (c. 600 B.C.E.): "All Cretans are liars."

In 1905 a French professor of mathematics from Dijon, Jules Richard, published a paradoxical definition of a number: "Consider the smallest number not definable in English in less than twenty words." But Richard had just defined this number in thirteen words! Richard's statement of this paradox was published in a journal with a large audience, the *Revue générale des sciences*.

This new contradiction in logic stimulated both Poincaré and Russell to find a solution by excluding difficulties stemming from "non-predicative definitions," called in common language "vicious circles." As Poincaré later commented about the axioms permitting non-predicative definitions such as Richard's, "The sheep-fold is well locked, but I am afraid the wolf is locked inside."

All these difficulties diminished the enthusiasm of Borel and Lebesgue for set theory. Borel and Russell met in Paris, but the contact was not warm. Discussions of these paradoxes went on between Poincaré and Russell for a few more years. But the worst was yet to come.

The Heidelberg Congress of 1904: The Fight Begins

The next Congress of Mathematicians after the Paris one in 1900 was held at Heidelberg in 1904. And here a dramatic event occurred. With Georg Cantor sitting in the audience with his wife and daughters, the Hungarian mathematician Julius König announced that the Continuum Hypothesis was wrong and that the cardinal of the Continuum was not an aleph. Cantor was deeply downcast even though Cantor, Bernstein, and König himself soon found a mistake in the proof of this statement.

On September 26, 1904, the German mathematician Ernst Zermelo, a student of Max Planck in statistical physics who had turned to the foundations of mathematics, wrote a letter to David Hilbert telling him that he had solved the problem of the Continuum. His proof made use of what would soon become well known to mathematicians as the "Axiom of Choice": "For any family of non-empty sets there exists a correspondence that associates to each of these sets one of its elements." That is, given a family of non-empty sets, one can "choose simultaneously" an element in each of them. In particular, if a set is non-empty one can choose one specific element in it. Hilbert decided that this letter deserved a wide audience, and almost immediately published it in his journal *Mathematische Annalen*.[18] The article, written in an unusual leisurely style, caused a sensation. As Lebesgue observed, "Zermelo arrived and the fight began." Indeed,

The French Trio

Jacques Hadamard.

Zermelo's proclamation stimulated a debate that lasted for more than ten years. The first reaction was Borel's, which Hilbert published in December 1904: Borel objected to the Axiom of Choice because "such reasoning does not belong to mathematics."

An exchange of five letters occurred in 1905 among four French mathematicians—Borel, Baire, Lebesgue, and Hadamard.[19] In these published letters Borel, Baire, and Lebesgue all rejected Zermelo's Axiom of Choice. Only Hadamard did not completely oppose it. He took a very personal approach, saying that "the question of what is a correspondence that can be *described* is a matter of *psychology* and relates to a property of the mind outside the domain of mathematics."[20] Needless to say, this view only increased the critics' hostility. The implicit question was: Is mathematics a house built on sand, on the shaky foundations of psychology and philosophy?

By emphasizing the importance of "selecting a correspondence,"

{ 57 }

Zermelo had raised the questions: "What does it mean to choose?" "Is it possible to make an infinity of choices?" In his Axiom of Choice Zermelo said nothing about how one is supposed to choose, or how the element to be chosen is to be specified.

In contrast to Hadamard, Lebesgue was trying to separate mathematics from psychology, but, like Hadamard, he spurned the idea of infinite choices:

> To define a set is to analyze objects in a bag C; we know only that the objects in the bag C have a property B in common that others do not have. One does not even know how to distinguish them.

Lebesgue also recognized the central issue in the debate when he asked, "Can we convince ourselves of the existence of a mathematical object without defining it? To define always means *naming* a characteristic property of what is being defined" (emphasis added). In Lebesgue's use of the term "naming" *(ensembles nommés)*, we catch a hint of the importance of the concept later to the Russian Name Worshippers. The ontological status of mathematical objects was at stake.

It is striking that Borel, who had shown courage in his recent work by using the transfinites of set theory, now in the fight occasioned by Zermelo's axiom contradicted what was implicit in his earlier views. He appeared to be losing his enthusiasm for the furthest extensions of set theory, probably because of the various paradoxes and difficulties that mathematicians had been encountering.

Lebesgue was insisting on the issue of sets known to be non-empty, but such that it is impossible to find explicitly any element, as in the case of "normal numbers." Normal numbers are numbers with decimal expansion exhibiting perfect randomness. Although the existence of such numbers can easily be proved, explicit namings of even one of them have been very difficult to obtain. The example is important because it was the occasion for Borel to introduce new ideas in probability theory.[21]

Mathematicians would have to wait for the rise of Nikolai Luzin's

The French Trio

Moscow School to get profound results on the naming questions raised by Borel and Lebesgue. For example, functions of higher Baire class were not given until more than twenty years later by a brilliant female member of the Russian group Lusitania, Ludmila V. Keldysh (1904–1976).

All these difficulties over the years made Lebesgue and Borel retreat even further, rejecting not only the Axiom of Choice but also the use of transfinite numbers. As late as 1908 Borel still opposed the use of non-denumerable infinities, and even denumerable infinities if they were not effectively constructed step by step.[22]

New attacks continued to come from the steadfast opponents of set theory, such as Picard, who in 1909 gave a humorous summary of the situation:

> These speculations about infinity are a completely new chapter in the history of mathematics of recent years, but it is necessary to recognize that this chapter does not escape paradoxes. Thus, one can define certain numbers that belong, and at the same time do not belong, to specific sets. All problems of this type are caused by a lack of agreement on what existence means. Some believers in set theory are scholastics who would have loved to discuss the proofs of the existence of God with Saint Anselme and his opponent Gaunilon, the monk of Noirmoutiers.[23]

Picard was obviously referring here to Richard's paradox and to the classical ontological debate about nominalism, but at the same time he was raising the issue of religion in order to discredit set theory. In contrast, some of the Russian mathematicians we will encounter later would appeal to religion to strengthen set theory.

Picard was not the only one to use irony in regard to serious matters. Even much later Lebesgue would remember with humor and nostalgia this rich period in his life. In 1938 Lebesgue was given an honorary degree in Lwow, and he was taken to the coffee shop where the famous Polish mathematician Stephen Banach used to work. The waiter handed him a menu with long descriptions in Polish. Le-

Charles-Émile Picard.

besgue glanced at it and answered, "Thank you, I only eat well-defined objects." The mathematician who was accompanying him immediately added, "You certainly are right to eat only meals defined by a finite number of words!"

The exchange of five letters about the Axiom of Choice was important for several reasons:

- It represented a real turn in the development of mathematics, whose foundation was at stake. A little later the German answer to the difficulties of set theory would lead to the birth of the axiomatic method developed by the Hilbert School, and, later, the Bourbaki group in France.
- It is a unique example of the close intertwining of personalities in a creative process and the mixing of mathematical, philosophical, and psychological issues.

The French Trio

- We have, a century later, partial answers to the problems that were raised, in particular the famous independence results of Gödel and Cohen. Still, all is not resolved, and the word "End" is not yet written.

Although all four of the French mathematicians who participated in the exchange about the Axiom of Choice had a common empirical approach to the problems being discussed, one can make a distinction, which grew with time, between Lebesgue and Borel. Borel's well-defined objects needed to be explicitly computable, and in this way Borel was not far from forecasting the future theory of computability (with its key notion of recursivity, introduced more than twenty years later). Lebesgue, on the other hand, tried to give a less restrictive limitation: his notion of "nameable object" *(objet nommable)*, introduced for the first time in 1904, refers to an object for which a characteristic property has been named. (We will see Russian developments of this idea later.) Lebesgue did not always ask for the Borel property of an explicit way of computing the object.[24] This attitude on the part of Lesbegue (similar to the abstract construction of his Integral) contrasted with Borel's insistence on explicit definition, and was a step in the direction of the axiomatic building of mathematics later constructed by German mathematicians. This explains partly why it was Lebesgue, not Borel, who wrote the introduction to Luzin's master work in French in 1930.

Borel and Lebesgue understood that this difference existed between them. In a 1919 article entitled "On Analytical Definitions and on the Illusion of the Transfinite," Borel admitted that his point of view was more restrictive than Lebesgue's. He further confessed that Lebesgue's approach might be more useful if one were not as critical as he was on the "illusions of the transfinite."

Borel came to realize that set theory was not for him anymore because his (Cartesian) sensual realism could not cope with such abstractness. But he was not a man to abandon the fight easily, and he had strong regrets about ceasing creative work in such a fascinating domain. Borel kept up with set theory and the theory of functions by writing numerous articles and books (with many editions and new introductions for each). For example, at the International Congress of Philosophy of Science in 1951, just a few months before his death, he gave a talk on "definition in mathematics." But years earlier he had become interested in the applications of his measure theory and Lebesgue's integration theory to the existence of normal numbers, and later he extended the applications to probability theory. In 1909 he wrote to Camille, "Not having any more the strength for high mathematics, I will go safely to work in probability and statistics following your uncle Bertrand.[25] It is not much compared to my earlier works in mathematics, but it is useful!" (Or as Borel colorfully expressed it in French, "Je vais pantoufler dans les probabilités.")[26]

Borel got involved in numerous other activities. He promoted, along with many other mathematicians, the reform of French education which took place in 1902. In 1904 he participated in a conference at the "Musée pédagogique" where he explained clearly his views on mathematics and mathematical education: "One must look for all occasions . . . for our pupils to realize that mathematics is not pure abstraction." In 1905, typically adapting to circumstances and mixing his personal interests with his love for Camille, he created a journal with her, *La Revue du Mois*, which lasted until a few years after World War I. In the first issue of January 1906 there was an article by Vito Volterra on the use of mathematics in the biological and social sciences. Borel added a short comment on the current debates about set theory, mentioning the remarkable recent work of Lebesgue and his "named sets."

Lebesgue, however, was still attracted to the geometric mysteries of the Continuum, and this fascination may have pushed him into a very elementary mistake, one that had important consequences. This mistake left an opening for the Russian mathematicians Suslin and Luzin, who welcomed set theory, to correct the error twelve years later.

The French mathematicians of this period did not want to mix

The French Trio

psychology or philosophy (not to mention religion) with mathematics, but instead wanted to restrict mathematical notions to those for which a clear definition as well as a clear "representation in the mind" could be found. This skepticism on the part of the French mathematicians, and their strong opposition to the new mathematics, prevented them from going further in set theory.

But still in 1909 Borel wrote: "From the day set theory stops being metaphysical and becomes practical, the new ideas may produce a flowering of beautiful results. . . . Maybe from this profusion of formal logic, which appears as a construction without any basis, one day some useful idea will come."[27]

The French reluctance to continue with set theory did have a positive result: they forced the German school (headed by Hilbert) to develop metamathematics, which produced the axiomatic method. They also, as we will see, stimulated the Russians to new creativity.

The events discussed here also had personal consequences. Baire experienced increasing mental difficulties and eventually, living alone in a hotel on Lake Leman, committed suicide. Borel abandoned "high mathematics" and even confessed to Paul Valéry in 1924 that he had become frightened of the mental consequences of research on set theory, referring to "poor Baire."

In fact, a number of scientists in other fields also suffered psychologically. Paul Langevin, the brilliant physicist and friend of Borel, spent many years in a poor mental state as a result of the pressure he was experiencing from his wife, who wanted him to leave pure research for private industry. But set theory presented particular problems. Even some of the Russians, who were more open to set theory, were not immune to these disruptions. Pavel Alexandrov confessed to the Hungarian mathematician George Pólya (who told it to Jean Dieudonné) that after a year of working on the Continuum Hypothesis, he became seriously worried about his mental equilibrium.

After 1917 Lebesgue began to feud with Borel. The pretext was priority on the birth of measure theory, but Borel's social life and activities were not all pleasing to Lebesgue, even though Camille tried to keep the two on good terms. Lebesgue was furious about the fact that he was placed under Borel's command when the latter was chief of scientific defense activities during World War I. The last letter of Lebesgue to Borel, dated December 21, 1917, is a beautiful and sad testimony to the death of what should have been an eternal friendship between two exceptional men. Lebesgue wrote:

> I don't have the courage to rebuff your proposals. I told you, I don't have the same confidence in you as I used to. I don't believe in words anymore. . . . For the moment any kind of relation going further than plain comradeship [*la banalité de la camaraderie*] would just be hypocrisy. I would not be having lunch with you but with some old memories. I think this letter will bring some sorrow to you and I keep too much hidden friendship for you not to be sorry myself.

All three members of the French trio eventually confronted an intellectual abyss before which they came to a halt. But each member of the trio reacted to the abyss differently. Borel abandoned the field but was not psychologically damaged by his change in focus. His world was a rich one, with many attractions in addition to mathematics: his love for his wife, politics, culture. Lebesgue was less flexible and generous, and in his frustration he became somewhat sour. The "malicious smile" that Camille Marbo had noticed in his youth became a form of heightened criticism of his colleagues, even those who deeply admired him, like Borel. Baire, frustrated both by his lack of professional recognition and his inability to cross the abyss, became more and more deeply depressed.

Thus the story of set theory in France has many intertwined factors: personal characteristics (including much intellectual creativity), attitudes toward philosophy and metaphysics, family situations, and politics. In Russia a similar complex mix was at work, with great intellectual creativity and even more personal sadness, and with stronger religious and political influences. But the most important differ-

The French Trio

ence between the French and Russian analysts of set theory is that the link between mathematics and metaphysics that the French tried to avoid was a connection that the Russians welcomed. Indeed, one could say that in the hands of the Russian Trio, metaphysics became mysticism.

4

The Russian Trio: Egorov, Luzin, Florensky

> "I felt as if I had leaned on a pillar . . . I owe my interest in life to you."
> —Nikolai Luzin to Father Pavel Florensky, July 1908

THE LIFE STORIES of Dmitri Egorov, Nikolai Luzin, and Pavel Florensky reveal much about their attitudes toward intellectual and religious issues, as does the cultural and political milieu from which they came. We will begin with that milieu, focusing on several of their Russian predecessors in mathematics and their ways of thinking.

Russian mathematicians in the late nineteenth and early twentieth centuries viewed their work as closely connected with philosophical, religious, and ideological issues. In that respect they differed from most of their French colleagues. Russian mathematicians tended to see knowledge as an interlinked, united whole, and thought that anything new that arose in one realm had effects elsewhere, perhaps even everywhere. As mathematicians, they felt obligated to relate their work to the larger world of knowledge and belief. Such views could also be found in western Europe, of course (Quetelet and Buckle are examples), but in late-nineteenth-century Russia they were particularly strong; this was a place where controversies over monarchy,

The Russian Trio

religion, determinism, free will, and Marxism became very heated and often entered into discussions in scientific fields. This tradition set the stage for the different reception of set theory in Russia as opposed to France.

A professor of mathematics at Moscow University who became involved in such debates even before the advent of set theory in Russia was Nikolai Vasilievich Bugaev (1837–1903). Bugaev would end up as a teacher of all three members of the Russian trio of mathematicians. He was also the father of Andrei Bely, the symbolist poet, who would study mathematics under Egorov and whose beliefs in "the magic of words" and the importance of "naming" became well known.

After graduating with a degree in mathematics and physics from Moscow University, Nikolai Bugaev did doctoral work in mathematics in Berlin and Paris. He then returned to Moscow and taught at the university for the rest of his life. He was an early member of the Moscow Mathematical Society, founded in 1864, one of the oldest mathematical societies in the world. His research concentrated on analysis and number theory, and he gradually developed a strong interest in the theory of discontinuous functions, which he called "arithmology." His attraction to discontinuous functions was in part ideological and religious.

Bugaev lived with his beautiful and rich wife, Aleksandra Dmitrevna, in an apartment on Arbat Street, the center of intellectual and artistic life in Moscow. This was the area where both Egorov and Luzin would eventually reside, all within a few blocks of each other and not far from Moscow University. The Bugaev apartment, located near a building where the great poet Alexander Puskhin had once lived, was a gathering place for socially prominent intellectuals (just as the Borel apartment was in Paris). Bugaev was also a member of the Russian Psychological Society, and in 1889 he published in the journal of that society a paper entitled "On Freedom of Will" in which he praised freedom of will as the most human of all characteristics. Here the mathematician was stepping directly into the world of philosophy and ideology. Bugaev was eager to defend free will because he saw it as the foundation of the autonomy of persons, and as closely connected with law, morality, education, and sociability. He believed free will was threatened by the deterministic philosophies that were prevalent in nineteenth-century Europe.

Eight years later, at the First International Congress of Mathematicians in Zurich in 1897, Bugaev connected his defense of freedom of will to mathematics itself. Discontinuous functions, so often regarded by mathematicians as frightful and repellent (the French mathematician Hermite called them "monsters"), were actually, said Bugaev, beautiful and morally strengthening because they freed human beings from "fatalism." "Discontinuity," Bugaev told his fellow mathematicians, is a "manifestation of independent individuality and autonomy. Discontinuity intervenes in questions of final causes and ethical and aesthetic problems."[1]

The importance of mathematical discontinuity would later be a central topic for the Name Worshipper Pavel Florensky while he was still a mathematics student; it was connected in his mind to the act of "renaming." Florensky's interest undoubtedly derived in part from the views of Bugaev, who was his mathematics professor at the university. Florensky described Bugaev's lectures in a way that clearly revealed the connections his professor saw between mathematics and social questions:

> We have a truly fine professor here in Bugaev, who is rather well known by his works. He intersperses in his lectures sharp comments, aphorisms, comparisons, and he gets into psychology, philosophy, ethics, but does all this in such an appropriate way that one is able to understand his explanations more clearly.[2]

Another example of the linking of mathematics and social questions by Russian mathematicians was a debate about freedom of will that took place in the years 1892–1903 between P. A. Nekrasov in Moscow and A. A. Markov in St. Petersburg. Markov and Nekrasov shared the view of many Russian mathematicians that mathematics impinged on ideology, but they drew opposite conclusions about

The Russian Trio

what those effects were: Nekrasov was motivated by religion, Markov by the French rationalist tradition.

Nekrasov (1853–1924) was a man devoted to the tsarist autocracy and the Russian Orthodox Church. He believed that ideas like determinism, atheism, and Marxism were closely linked, and he criticized those concepts in the name of Christianity and free will. Nekrasov wrote books and articles in which mathematics, theology, and philosophy were intertwined.

Markov (1856–1922), on the other hand, was an atheist and a strong critic of the Orthodox Church and the tsarist government (Nekrasov exaggeratedly called him a Marxist). The Nekrasov-Markov dispute helped strengthen the common opinion that the St. Petersburg School of Mathematics (Markov's home) was secular and philosophically "positivistic" if not "materialistic," a seat of liberal democracy and anti-monarchism; in contrast, Moscow mathematicians such as Nekrasov were often seen as more religious, more favorable to monarchy, and prone to philosophical "idealism."

The topic of the debate between the two was the explanation of statistical regularities. The Law of Large Numbers, when applied to non-human situations such as the drawing of balls from an urn containing both black and white balls, points to the non-controversial conclusion that the larger the sample of balls one takes from the urn, the closer the proportion of black to white found in the sample will be to the proportion contained in the urn itself. (Jacob Bernoulli did important work on this mathematical relationship in the early eighteenth century.) However, in the nineteenth century a man often called the founder of modern statistics, L. A. J. Quetelet (1796–1874), applied the Law of Large Numbers to human beings on such topics as the ages at which men and women marry. The year-to-year consistency of these statistics (when large numbers are involved) led him to the conclusion that the role of free will seems to "wash out," and that human behavior can be described in terms of statistical regularities so predictable that they approach the strength of laws in physics. Quetelet, in fact, wrote a book with the meaningful title *Social Physics*. Arguments like Quetelet's were regarded as threatening by some

{ 69 }

defenders of religion, since the prevailing opinion in the Judeo-Christian tradition favors free will (although the historical record of theological debate on this issue is variegated and complex). Quetelet had raised the possibility that the alleged free will of human beings is merely a chimera.

Nekrasov was deeply disturbed by terms like "social physics" and attempted to rescue a concept of free will by mathematical examination. He noted that the assumption behind the Law of Large Numbers (when applied to such situations as an urn with both black and white balls in it) is the *independence* of successive experiments. (Balls that are extracted from the urn must be returned to it, so that each extraction is unaffected by any previous ones.) But Nekrasov thought that among human beings the situation was different, and he developed the concept of "pairwise independent" (rather than "mutually independent") random variables. He asserted that pairwise independence was not only sufficient but necessary for the Law of Large Numbers to hold. For him, free will was tantamount to pairwise independence.

Markov, seeing the ideological implications of the discussion, was offended by Nekrasov's conclusions. He literally changed his research direction to oppose Nekrasov. In a letter to a colleague written on November 6, 1910, Markov explained his motivations for going in a new direction:

> The unique service of P. A. Nekrasov, in my opinion, is namely this: he brings out sharply his delusion, shared, I believe, by many, that independence is a necessary condition for the law of large numbers. This circumstance prompted me to explain, in a series of articles, that the law of large numbers ... can apply also to dependent variables. In this way a construction of a highly general character was arrived at, which P. A. Nekrasov cannot even dream about.
>
> I considered variables connected in a simple chain, and from this came the idea of the possibility of extending the limit theorems of the calculus of probability also to a complex chain.[3]

The Russian Trio

Thus "Markov chains" were born, one of the significant concepts of modern mathematics. A Markov chain is a sequence of random variables with the "Markov property"—namely that given the present-state, the future and past states are independent. Examples of Markov chains can be found in any game, like Monopoly, whose moves are determined entirely by dice. Such games contrast with card games like poker or blackjack, where the cards represent a "memory," since what is displayed at any moment is dependent on past moves.

One of Bugaev's students, Dmitri Fedorovich Egorov, would develop a deep interest in the relationship of religion and mathematics.[4] Born in 1869 in Moscow, he spent his entire life in that city, with the exceptions of study abroad in Europe and his imprisonment and death in the city of Kazan. Egorov's father, Fedor, was a mathematics teacher and the director of the Moscow Teachers' Institute, a school which gave a three-year education to future secondary teachers in the city. Fedor Egorov taught algebra and geometry there, and he imparted to his son Dmitri a love for the same subjects.

Dmitri graduated from Moscow School No. 6 with a gold medal and entered the physics-mathematics department of Moscow University, where his main professor was Bugaev. It was from him that Egorov picked up his strong interest in discontinuous functions. For a while Egorov was also interested in Bugaev's "arithmology," but he soon abandoned that topic for differential geometry, a subject on which he wrote a student paper. In 1894 he became a *privat-dotsent* (unestablished university lecturer) at Moscow University. In 1899 he defended his master's dissertation, followed in 1901 by his doctoral dissertation, with the title of "Ob odnom klasse ortogonal'nykh system" ("Concerning One Class of Orthogonal Systems"). This became a classic paper which led the French mathematician Gaston Darboux to name a type of topological surfaces in his honor as "Egorov surfaces." Egorov's reputation as a leading specialist in differential geometry was established.

In 1902–1903 Egorov went to Europe, where, in Berlin, Göttingen, and Paris, he attended lectures given by some of the best-known mathematicians of the day, including Frobenius, Poincaré, Darboux, Hadamard, Lebesgue, Klein, Hilbert, and Minkowski. He was particularly attracted to Lebesgue's work on the theory of functions. Returning to Moscow University, where in 1904 he became a professor, Egorov launched a brilliant career as a teacher who influenced a whole generation of mathematicians. Together with B. K. Mlodzeevsky (who had introduced set theory to the university in 1900–1901), Egorov taught his students about the latest developments in European mathematics and invited them to participate in its further elaboration.

Egorov was a very reserved and modest man, so much so that it would be easy to believe that he lived only for mathematics. Unlike his teacher Bugaev, who often wrote and lectured on topics like free will and psychology, Egorov considered such extra-scientific publications and lectures to be unwise, perhaps even improper, even though he had his own views on religious and philosophical topics. He was a scholar, not a publicist. His publications do not reveal any evidence of the "inner Egorov"—indications of the motivations that were so clear in many of his predecessors like Bugaev, Markov, and Nekrasov. However, a close study of his life shows that Egorov was a man of deep passions, religious commitments, cultural identity, and political preference. As Sergei Demidov, a leading Russian historian of mathematics, wrote in the post-Soviet period, Egorov "thought that the opinions and beliefs of a person (including his religious views) belonged to an intimate human sphere and were not a subject of discussion." That sphere was fundamentally important to Egorov, however, and one can see that it affected his mathematics as well as his personal life.

Egorov's social views are easier to detect in his actions than in his writings. He attended church regularly, and his interest in links between religion and mathematics steadily grew, culminating in friendships with many priests and in his participation in discussion groups with church people, philosophers, and scientists. In politics also his commitments must be sought in his behavior, not in his publications. In 1903 there was a pogrom against Jews in Kishinev which resulted in violent deaths. Quite a few intellectuals in Moscow and St. Petersburg protested, including scientists. Some wrote articles decrying the horrible events. Egorov did not write articles, but if one looks at the petitions signed by some of these intellectuals protesting the pogrom, along with names like Leo Tolstoy and Vladimir Vernadsky one will find the name "D. F. Egorov."

The influence of the French mathematicians he had just visited in Paris played a role in Egorov's protest against the pogroms. As we have seen, France when he visited there in 1902–1903 was in an uproar over the Dreyfus Affair. Several of Egorov's French colleagues, including Lebesgue, Poincaré, and Borel, were among the "Dreyfusards," the intellectuals who defended Dreyfus. The example of their protests made an impression on Egorov, who by his nature was reluctant to engage in political action. It was shortly after his return from France that Egorov, in a rare moment of public demonstration outside his field, appended his signature to the petition denouncing the pogrom in Kishinev.

Egorov's name is not associated with the radicals who played a role in the revolution of 1905, a time when many intellectuals criticized the tsarist autocracy. Egorov was a defender of a moderate, constitutional monarchy, but his letters to Vladimir Vernadsky in May and June of 1905 show that he actively supported the movement for reforms in university life, working toward achieving a more responsive university administration.[5]

Egorov was closely connected with the richness and political complexity of European culture. Democratic ideals were of course present in European thought at the turn of the century, but they were still not fully developed. Germany, after all, where Egorov studied, and whose universities he so obviously admired, was still a monarchy—a constitutional one, to be sure, with political parties and elections, but still far from democratic. France, where Egorov also studied, was, on the other hand, clearly a republic, but one with a tumultuous political life that probably seemed somewhat bewildering to Egorov. But Europe was a powerful cultural magnet for Dmitri Egorov and his family, and it shaped their social and public lives.

Egorov's wife, Anna Ivanovich Grzhimali, was the daughter of Ivan Grzhimali, one of the most famous violinists in Russia who performed all over Europe. Dmitri's marriage to Anna placed the young couple in the midst of the leading cultural and social elite of Moscow, just as his professor Bugaev and his wife had been. Dmitri and Anna frequently attended dinners and parties in the Grzhimali apartment, which was in a building of the Moscow Conservatory where Ivan Grzhimali was a professor. The apartment was spacious and luxurious, with 16-foot ceilings. There the Grzhimalis, often together with Dmitri and Anna Egorov, entertained the social and cultural leaders of Russia, including Tchaikovsky, Chaliapin, Rachmaninoff, Ilya Repin, and many others. The Grzhimali home was an open one, and Ivan was known to everyone.

Ivan Grzhimali was born in Prague and moved easily among the European musical community. His father-in-law, only eight years older than he, was Ferdinand Laub, also a violinist famous throughout Europe. Laub was Jewish, so Dmitri Egorov's wife Anna (Laub's granddaughter) was part Jewish. Laub and Ivan Grzhimali were close friends. They both enjoyed the company of musicians in western Europe and Russia, including Liszt and Tchaikovsky.

The Grzhimalis had two daughters, Dmitri Egorov's future wife Anna (known to her close friends as Aida) and Natalya. Both Anna and Natalya were musically gifted, not surprisingly given the talent of their father. But the gossips of Moscow said that it was not necessary to assume that Ivan Grzhimali was Natalya's father in order to explain her musical brilliance; according to some of them, her real father was Liszt, who allegedly was involved with Natalya's mother during one of the Grzhimalis' frequent concert tours in Europe.

Both Anna and Natalya received their musical educations at their father's institution, the Moscow Conservatory. Anna became an accomplished amateur pianist and singer, while Natalya, who never married, became a professional pianist and lived for many years with Dmitri and Anna Egorov. The Egorovs did not have children, and the gossips of Moscow, never tiring of their stories, maintained that

The Russian Trio

Dmitri and Anna never had sexual relations, linking this curious fact to Dmitri's unusual religious beliefs—not an entirely credible rumor, since even priests in the Orthodox faith can marry and have children, as did Dmitri's friend Pavel Florensky, who shared his Name Worshipping beliefs.

In the Egorov apartment, located on Boris and Gleb Street in the prestigious Arbat region of Moscow, music was often in the air. Two pianos competed with each other, a grand piano for Natalya and a smaller one for Anna. Natalya frequently gave lessons there for young piano students (including one still living today, who described the apartment). Sometimes her father Ivan came to the apartment for social events, where he would play for the guests on his rare Stradivarius violin. Just as Émile Borel's marriage to Marguerite Appell helped situate him in the center of Parisian intellectual and cultural life, Dmitri Egorov's marriage to Anna Grzhimali put him in a similar rarefied milieu in Moscow. When the two mathematicians met, their similar social environments helped cement their relationship.

The social world in which the Egorovs and Grzhimalis moved is beautifully reproduced in perhaps the best-known poem of the symbolist poet Andrey Bely, who knew that milieu well. The poem is entitled "The First Encounter" ("Pervoe svidanie"); in one of its major sections Bely describes the cultural and intellectual leaders of Moscow at one of the musical concerts held at the Nobles' Club. The women, wearing fashionable boas, are dressed in stylish dresses from the best designers of Moscow, such as Minangois and Lamanova (who had a "salon de couture" until the Revolution). The latest fashion is that of the "cloche," a belted skirt ten yards wide at the hem, which, particularly if made of silk, made a rustling sound as the women walked. Bely observes:

> I see these ladies—
> In boas—stout and noble;
> And—others: feathered, ardent ladies
> Beautiful ladies in their stylish capes . . .

Grzhimali is described as the concertmaster of the evening, obviously the object of much admiration. In one stanza Bely rhapsodizes:

> My mystic panoramas
> Are twirled up by the flowing bow,
> The bow both weepy and familiar—
> Of Grzhimali's violin.

Egorov, on the other hand, is seen by Bely in very different terms, those of modesty. As he emerges from the concert hall he "sheepishly bends downward" before the others in the crowd, not asserting the eminence that he, too, could justly claim.[6] After all, he was prominent in his field, a professor of mathematics at Moscow University, known among European mathematicians for his achievements, and also closely related, through his wife, to one of the stars of the evening, Ivan Grzhimali. But Egorov was a private and humble man whose very posture portrayed his personality.

Although Egorov was taciturn and immensely polite, he was far from neutral on principled questions. He once said that when he was at a meeting and heard something with which he disagreed, he felt obligated to express this disagreement. This unusual honesty did not usually cause great difficulties in mathematics seminars—where people were expected to disagree—but in social gatherings it could lead to trouble, and in the political environment of Soviet Russia after 1917 it could be dangerous.

Egorov was deeply religious. The mathematician N. M. Beskin wrote, "On getting to know Egorov, one was struck with his religiosity. At his home I saw priests with whom he was deeply respectful, kissing their hands on meeting. On his desk, along with mathematical books, was the Bible. He alternated reading mathematical literature with the reading of theological literature for relaxation."[7] In June 1914 Egorov wrote to his colleague and former star student Nikolai Luzin that he had just been reading Pavel Florensky's dissertation and found "much of interest in it." This was the deeply religious dissertation in which Florensky praised mathematical discon-

The Russian Trio

tinuity (just as Bugaev had done in his day) as an escape from determinism and fatalism.

Egorov's student Nikolai Luzin (1883–1950) was without question one of the major Russian mathematicians of the twentieth century, someone who had an international influence through the Moscow Mathematical School, which he founded along with Egorov. Luzin's life story is both fascinating and troubling, and there are still some questions and debates about it. But enough is now known so that we can have a fairly good picture of his life.[8]

The sources still differ over several basic facts, starting with his place of birth—some say that he was born in Irkutsk, others in Tomsk. Nina Bari, who knew Luzin almost as well as anyone and was close to him for over thirty years, stated after his death in 1950 that he was born in Tomsk and moved briefly to Irkutsk when he was still a young boy.

Luzin's paternal grandfather was a serf on one of Count Stroganov's estates near Tomsk. His father was a businessman in the Tomsk area and his mother was a Buryat, a people of the Buddhist faith, from the Baikal region. Luzin's initial education was in a private school in Tomsk. At the age of eight he went to a gymnasium in Tomsk, then to a school in Irkutsk for one year (where his father was sent on business), and later back to Tomsk.

During these early years of education Luzin did not display any particular talent or interest in mathematics; instead, he was attracted to romantic literature and philosophy. He actually feared mathematics, which was taught to him as something that had to be mechanically memorized—a system of methods (addition, subtraction, division) or a collection of standard theorems or equations. Since Luzin had a poor memory, he looked upon his early mathematics teachers more as torturers than as helpers. He did badly with history for the same reason: he could not remember names and dates very well.

Luzin's grades sank lower and lower, to the point where his desperate father hired a tutor for him. Fortunately, this young man, a student at the Tomsk Polytechnical Institute, took an entirely different approach from Luzin's earlier teachers. Mathematics to him was not a subject for routine memorization, but instead a method of inquiry based on reasoning and imagination. This new view of mathematics inspired the young Luzin, and he lost his fear of the subject. In fact, in a few years he was the star mathematics student in his Tomsk gymnasium.

After completing his secondary schooling, Luzin in 1901 gained admission to the physics-mathematics department of Moscow University, a transition that was made easier by the fact that his father sold his business in Tomsk and moved to Moscow. At first Luzin lived in the new family home in Moscow, but then his father had a series of financial reverses (partly as a result of gambling) and had to sell their house.

Luzin found a room, together with his friend V. A. Kostitsin, in a house owned by the widow of a physician named Mikhail Malygin. This new accommodation affected Luzin in several ways. His friend was involved in revolutionary activities and, according to later stories, even hid some explosives in the same room which he shared with Luzin. Although Luzin did not become a revolutionary, he was influenced by the conversations with his friend about the nature of the autocratic regime under which they both lived. By not reporting his friend's explosives to the authorities, he was technically complicit in revolutionary activities. Luzin was, as a young man, more radical than his teacher Egorov and believed, in at least an abstract way, that a revolution would be a positive development in Russia. But Luzin was an extremely sensitive young man and was far too nervous and indecisive to make a good revolutionary himself. Nonetheless, he was critical of the autocracy and its tight link with Russian Orthodoxy. Like many Russian intellectuals, he saw science, secularism, and philosophical materialism as liberating ideas. Eventually his friend left and went into hiding, evidently taking the explosives with him.

The rented room had another effect on Luzin's personal life. The widow who owned the house had a daughter with whom Luzin became friendly, and a few years later he would marry the widow's daughter, Nadezhda Mikhailovna Malygina. But before that happened Luzin experienced a deep crisis that changed his life.

Luzin's initial intention at the university was to acquire a mathematics education that would permit him eventually to become an engineer. But he soon fell under the influence of mathematics professors like Bugaev and Egorov, who introduced him to the exciting developments in mathematics coming from Europe, especially Germany and France. They portrayed the field as one of creativity in which there were many alluring secrets. Once again, as in Tomsk, stimulating teachers deepened Luzin's mathematical interests. In fact, Luzin was captivated. He was much more drawn to questions about the foundations of mathematics, number and set theory than he was in solving practical problems that engineers were interested in, and he gave up on his earlier engineering ambitions. In later years this interest in foundations problems went so far that Luzin occasionally bragged that he "never solved equations" anymore.

At Moscow University Luzin was influenced not just by his teachers but also by his fellow students, who debated politics, philosophy, and mathematics with him. Among these mathematics students were Pavel Florensky and Boris Bugaev (also known as Andrey Bely, the son of Professor Bugaev). Both of these men were, like Luzin, destined for fame. Pavel Florensky was deeply interested in religion and would eventually become a priest and a renowned polymath. Several years after they all graduated from the university (Bugaev in 1903, Florensky in 1904, Luzin in 1905), Florensky had an important influence on Luzin by turning him away from secularism to religious faith. The experience that Bely had with mathematics left a mark on his noted literary work, as did his later familiarity with Name Worshipping, the religious heresy that affected Egorov, Luzin, Florensky, and Bely in different ways. Egorov and Florensky actually became Name Worshippers, while Luzin and Bely were deeply influenced by this philosophical approach.

From 1905 to 1908 Luzin underwent a psychological crisis so severe that several times he contemplated suicide. One precipitating event was the unsuccessful revolution of 1905, an event that was sobering for many left-wing members of the intelligentsia who had talked romantically of their hopes for a revolution without comprehending the violence that would ensue. For these intellectuals, 1905 was a year of truth. Luzin was shaken not only by the shedding of blood but also by personally witnessing poverty and suffering.

The 1905 revolution brought carnage to some areas of Moscow, and Luzin witnessed acts of violence committed by both sides, the revolutionaries and the defenders of tsarism. Tsarist Cossacks on horseback wielded whips and sabers and charged the demonstrators, slashing and maiming dozens of people. Armed workers erected barricades and fired upon the Cossacks. Army troops engaged the revolutionaries in street-by-street fighting, and on several occasions used artillery against the armed workers. Luzin did not participate in any of this, but he observed a great deal of it. And in the aftermath, in the Aleksandrovsky Garden near the Kremlin and Moscow University, he saw prostitutes offering their bodies to passersby for kopecks, exposing their breasts to entice customers. The young, naive, and nervous Luzin went into shock at what he saw.

Luzin was so disillusioned and traumatized by these events that he lost the ability to continue his mathematical studies—or, more accurately, he lost all interest in mathematics. How could he study mathematics, he asked himself, when the world had gone mad? Earlier, he had embraced science, materialism, and secularism as the answers to Russia's problems. Now he doubted that these were any answers at all.

His teacher Egorov was known for his devotion to his students, and Luzin was his prize student. Concerned about Luzin's mental state, Egorov advised him to go to Paris, to meet there with mathematicians like Henri Lebesgue, Émile Borel, and Jacques Hadamard. He hoped that in Paris Luzin would regain his mental balance and that his French colleagues would re-ignite his interest in mathematics. Egorov managed to arrange a foreign fellowship for his troubled student.

The Russian Trio

Hotel Parisiana on the rue Tournefort in Paris, around 1915; the Parisiana is next to the pushcart.

On Egorov's recommendation, Luzin lived in Paris in the same small hotel where Egorov had stayed earlier, the Parisiana, on the rue Tournefort in the Fifth Arrondissement. A typical nineteenth-century Paris building, the Parisiana was located in the academic heart of the city, near the Sorbonne and many other elite educational and scientific institutions. It was not far from the École Normale Supérieure, intellectual home of Borel, Lebesgue, and Baire, and was also close to the Pantheon, the place where France honored its "great men." The history of the Pantheon was one that would later have resonance for both Egorov and Luzin. It had been constructed as a church, the Église Sainte-Geneviève, but during the French Revolution it was secularized, transformed into a place that celebrated the great men of literature and science, and deprived of its religious status. Little did Luzin and Egorov know, gazing at the dome of this magnificent building (still surmounted by a cross) or passing by it to enter the huge Bibliothèque Sainte-Geneviève on the north side of the square, that, as in France much earlier, a sweeping revolution would soon transform religious and intellectual institutions in Russia. The church at Moscow University, St. Tatiana the Martyr, would be similarly secularized, with consequences for both of them.

The proprietor of the Hotel Parisiana was a Monsieur Chamont, whose two little daughters were intrigued by the two Russian mathematicians who came frequently to stay in the hotel. Many years later, in the 1970s, the MIT mathematician Victor Guillemin would also stay there, and one of the Chamont daughters, now an old woman, told him that she remembered Egorov and Luzin, and recalled how hard-working and "pious" they were.

At first the move to Paris seemed to do nothing for Luzin. He was still totally without a purpose in life. He poured out his pain (in letters from both Moscow and Paris) to his friend Florensky, who was at the time in the Ecclesiastical Academy in Sergiev Posad. Luzin came to lean more and more heavily on Florensky in handling his mental crisis. In one of his letters Florensky agreed that "chaos and confusion" were now reigning in Russia. He told Luzin that one of the reasons for Russia's crisis was that so many of its brightest minds were attracted to agnosticism and atheism (just as Luzin had been). Florensky had himself, almost ten years earlier, undergone the transition from scientism to religion, and he sympathized with Luzin's plight, expressing the hope that Luzin would find a way to "the Source of all truth, to Truth itself."

In one of his letters to Florensky, Luzin wrote:

> It is painful for me to live . . .! Those worldviews which I earlier knew (materialistic worldviews) absolutely do not satisfy me. . . . Earlier I believed in materialism, but now I cannot live by it, and I have suffered, suffered, beyond end.[9]

Luzin went on to say,

> You found me a mere child at the University, knowing nothing. I don't know how it happened, but I cannot be satisfied any more with the analytic functions and Taylor series. . . . To see the misery of people, to see the torment of life. . . . —this is an

unbearable sight. . . . I cannot live by science alone. . . . I have nothing, no worldview, and no education.

And, most ominously, Luzin added, "If I do not find a path to seek the truth . . . I will not go on living."

Florensky supplied that path to truth, at least in Luzin's mind, but the full transition and conversion took a long time, perhaps two or three years. In the meantime Luzin returned to Russia, where he spent many days with Florensky in his monastery town outside Moscow, sometimes the entire summer. Again and again Luzin turned to Florensky for guidance, saying in one letter (March 14, 1908), "I want to see you, and only you." (This was the same year when Luzin married his former landlady's daughter.) In June 1908 Luzin read Florensky's thesis "On Religious Truth" (later published as a book, the English edition of which is still in print, called *The Pillar and Foundation of Truth;* Egorov read it in 1914 and corresponded excitedly with Luzin about it). The impact was profound. Luzin wrote his new wife, "I read it all at once in a single day—skipping a lot, but the impression was overwhelming. As I read it I was STUNNED the entire time by blows from a battering ram." He added, "This work is so valuable because it deals with the most fundamental questions of life." By July 1908 Luzin's religious conversion was complete, and he wrote Florensky that "I felt as if I had leaned on a pillar . . . I owe my interest in life to you." Luzin now found it possible to return to the study of mathematics, combining it with a deep interest in religious mysticism. His thoughts of suicide were behind him.

An interesting feature of this correspondence between Luzin and Florensky is that they sometimes called each other by different names. Luzin often referred to Florensky not by his correct name and patronymic, Pavel Alexandrovich, which he knew very well, but by a different name, Petr Afanasievich. Although several Russian historians of science have puzzled over this riddle, it has only recently become clear why Florensky and Luzin renamed themselves. In 2006, a century after its composition in 1906–1907, a book of Florensky's entitled *Holy Renaming* was published for the first time by the Church of

St. Tatiana the Martyr of Moscow University, an institution so important to Florensky, Egorov, Luzin, and the Name Worshippers.[10] In this work, based on research in eight or nine languages, Florensky maintained that an inspiration that came from mathematics had led him to a new appreciation of the importance of renaming persons who have religious conversions. Drawing on the views of his teacher Bugaev, who saw discontinuous functions as a "liberating" refutation of determinism, Florensky concluded that religious conversion was a redemptive discontinuity in spiritual life.

At the same time Florensky advanced the view that, following the ancient Christian custom of *adelphopoesis* (brother-making), male friends could be joined in chaste bonds of love. Moreover, according to Florensky, a way of signifying the existence of such a bond was to bestow different names on each other. Thus Florensky and Luzin bound themselves together in brotherly love on the basis of new names.

Although homosexuality would play an important role in the early years of the Moscow Mathematical School, there is no evidence whatever that Luzin and Florensky had an overt homosexual relationship, though they certainly had a very deep brotherly one. There is some evidence that this bond began earlier, in the spring of 1905, before Luzin graduated from Moscow University.[11] A further indication of their unusual relationship, which was intimate but at the same time distant and chaste, is that in their letters they referred to each other with the formal form of the pronoun "you" (*vy*). Neither Luzin nor Florensky was married at this time (Luzin married in 1908, Florensky in 1910); in later years their wives would become friends as well.

Although Nikolai Luzin remained married to Nadezhda Mikhailovna, the marriage had its problems. At first Luzin turned to his wife as an intellectual soulmate, discussing with her his consuming interest in set theory as a "mysterious area that envelops me deeper and deeper." He wrote Florensky that "my wife is also very interested and shares my commitment to the search for the profound truths of life." After a while, however, this attempt at intellectual partnership seems to have failed. In 1914 Luzin and his wife sepa-

The Russian Trio

Nikolai Luzin in 1917.

rated, and she then turned to Florensky for help, saying that his visits were invaluable for her husband. She obviously hoped Florensky would encourage her husband to return to her, and eventually he did. But she must have been surprised by the advice that Florensky gave her:

> Nikolai Nikolaevich is a very sweet and fine person; but in personal relationships he is not at all mature, especially in intuitively perceiving the hidden currents of life.... You will have to take the relationship in hand and create a family tone, simplicity. Instead, as I perceive it ... you have established the tone of an acquaintanceship rather than a family.[12]

Nadezhda Mikhailovna then evidently gave up all ideas of sharing the intellectual life of her husband and became a conventional housewife. And Nikolai became a somewhat conventional husband; he had a number of love affairs with other women, and even had a child by one of them.

We now turn to an examination of the life and work of the Russian priest and mathematician who had such an enormous influence on Luzin. Pavel Florensky was born in the remote town of Yevlakh in present-day Azerbaijan in the north Caucasus, where his father was a railroad engineer.[13] Pavel's father was Russian, from the city of Kostroma, and came from a family of Russian Orthodox priests. He was engaged in the construction of railroads in the Caucasus region that would convert Yevlakh into an important transportation hub. Florensky's mother (her maiden name was Saparova or Saparian) was from a cultured Armenian noble family who had moved nearby, to Georgia. Thus, by birth and by descent, Florensky was connected to the Caucasus, a region that remained important to him and to many of his later followers throughout their lives. The Caucasus was a place where Name Worshippers would later find a refuge from their persecutors, both in the late tsarist and Soviet periods.

As a youth Florensky lived in both Azerbaijan and Georgia, and often visited Armenia, his mother's home. The natural beauty and romantic qualities of the Caucasian mountains appealed to him, and he grew up and received his early education in this environment. In 1892 he entered a classical gymnasium in Tiflis, or Tblisi, capital of Georgia. His classmates there included the future philosopher V. F. Ern, who would later briefly join him in a radical Christian society that was suppressed by the tsarist authorities.

In his early life Florensky was not religious. His father, despite his descent from Orthodox priests, and his mother, of Armenian Christian origins, were members of the secular intelligentsia who had turned against religion in favor of science, which they considered a modernizing worldview. Florensky later wrote, "Educated in complete isolation from any religious notions and even from simple fairy tales, I regarded religion as something completely alien to me, and any religious lessons in the gymnasium drew from me only hostility and mockery."[14]

In 1899, however, when he was only seventeen years old, Florensky had a spiritual crisis that resulted in his religious conversion. This important event in Florensky's intellectual and spiritual life was a part of a trend among some Russian intellectuals at the end of the nineteenth century. A few of them revolted against the spread of Marxism and revolutionary ideology and began seeking alternatives in the Russian Orthodox Church, including some of its dissident strains. They were often critical of the control of the Church by the state, which dated to the reign of Peter the Great.

From the Caucasus Florensky went to Moscow University, where he entered the department of mathematics in 1900 and studied under Egorov and Bugaev. It was there that he first met Luzin and Bugaev's son Andrei (Bely), fellow mathematics students. In 1902 Florensky founded the student branch of the Moscow Mathematical Society, and in 1904 he appointed Luzin his successor as its secretary. It was at the university as an undergraduate that Florensky began his writings about mathematics and religion that would later influence both Egorov and Luzin.

In these writings Florensky defended the importance of the idea of "discontinuity" (a theme he undoubtedly picked up from his professor Bugaev), both in mathematics and in social behavior. Like many members of the Russian intelligentsia of this time, Florensky believed that all intellectual life is a connected entity, and that ideas in mathematics and philosophy could be extended to the social and moral realms.

Florensky was convinced that intellectually the nineteenth century, just ending, had been a disaster, and he wanted to identify and discredit what he saw as the "governing principle" of its calamitous effects. He saw that principle in the concept of "continuity," the belief that one could not make the transition from one point to another without passing through all the intermediate points. In contrast to this "false" principle of continuity Florensky proposed what he saw as its morally, even religiously, superior opposite: discontinuity. He realized of course that this was not a new topic, and that discussions of the antinomy of continuity/discontinuity were very old, dating back to the Greeks. However, Florensky believed that the problem had a particular relevance to the beginning of the twentieth century because, in his view, in the nineteenth century "the cementing idea of continuity brought everything together in one gigantic monolith."[15]

Florensky faulted his own field, mathematics, for creating this unfortunate monolith. Because of the strength of differential calculus, with its many practical applications, he maintained that mathematicians and philosophers tended to ignore those problems that could not be analyzed in this way—the essentially discontinuous phenomena. Only continuous functions were differentiable, so only those kinds of functions attracted attention. And this emphasis on the continuous, Florensky believed, affected many areas of thought outside mathematics. Differentiable functions were "deterministic," and emphasis on them led to what Florensky saw as an unhealthy determinism throughout political and philosophical thought in general, most clearly in Marxism.

Intellectual modes based on continuity, said Florensky, had spread to geology, in the uniformitarian ideas of Lyell, and to Darwin, in the concept of evolution through gradual small change. Both opposed "leaps" in natural development and postulated smooth, even transformations. Florensky believed that similar ideas had influenced many other fields, including psychology, sociology, and religion. He continued, "The idea of continuity, making these transitions, took possession of all disciplines from theology to mechanics, and it seemed that anyone who protested against its usurpations was a heretic."[16]

But now, said Florensky, the very field—mathematics—that was "guilty" of driving human thought into this blind alley was showing the way out of it. In the 1880s Georg Cantor, founder of set theory, had defined the "Continuum" as a mere "set of points" and in that

The Russian Trio

Pavel Florensky.

way had deprived the concept of continuity of its metaphysical, dogmatic power. Furthermore, Florensky's teacher Bugaev, at the First International Congress of Mathematicians in Zurich in 1897, had pointed out how concepts of discontinuity could be linked to freedom, aesthetics, and ethics when he wrote, "Discontinuity is a manifestation of independent individuality and autonomy." Now the road was open, Florensky believed, to restore discontinuity to its rightful place in one's worldview. He saw discontinuity reappearing almost everywhere—in mathematics (with the new interest in discontinuous functions, later so successfully developed by his friend Luzin), in biology (the concept of mutations), in molecular physics (electrons jumping between discrete rings around atoms), and in psychology (subliminal consciousness, creativity).

Upon his graduation from Moscow University in 1904, Florensky was offered a position as a graduate student to continue in mathematics. Luzin encouraged Florensky to accept, and even rushed to his apartment to urge him to do so.[17] However, Florensky declined the offer and instead entered the Theological Academy located in the monastery town of Sergiev Posad.

Luzin and Florensky, now following different professional goals, continued to correspond for many years. Although they had some differences of opinion (for quite a while Luzin insisted that infinity was only a "potential," while Florensky, like Cantor, considered it as "actual"), they remained close friends. The friendship lasted until the 1920s, when the new Soviet political atmosphere made such a link between Luzin, a prominent university professor, and Florensky, a priest, dangerous. Luzin even stopped going to church in the 1920s, although he resumed this practice after World War II.

Florensky, like Egorov and Luzin, was opposed to social activism, to the idea that intellectuals should engage in political activity. However, on March 12, 1906, Florensky delivered a sermon in which he deplored the execution of the naval officer Pyotr Shmidt, a participant in the unsuccessful 1905 revolution. Florensky did not share Shmidt's political opinions, but he opposed capital punishment. Students in the church who heard the sermon printed and circulated it on their own, with the result that Florensky was arrested by the tsarist police and held in jail for a week. While imprisoned Florensky wrote an essay entitled "On the elements of the aleph number system."[18]

After graduating from the Theological Academy in 1908 Florensky continued to teach there and lived with his family in Sergiev Posad, not far from the Trinity Monastery of St. Sergey. His home there was a simple log cottage with wooden fretwork above the windows and a central attic dormer, similar to many other traditional homes in Russian villages. (Today there is a plaque on the building indicating that it was his home for many years.) In this house Florensky amassed an enormous library in many languages on philosophy, science, and theology. It would remain there until confiscated by the Soviet secret police in 1933.

5

Russian Mathematics and Mysticism

"In the beginning was the Word, and the Word was with God, and the Word was God."
— *Gospel according to John, first verse*

"When I name an object with a word, I thereby assert its existence."
— *Andrei Bely, symbolist poet and former mathematics student of Dmitri Egorov, in his essay "The Magic of Words"*

"Nommer, c'est avoir individu" (to name is to have individuality).
— *Nikolai Luzin, leader of the Moscow School of Mathematics*

THROUGHOUT the entire history of mathematics, going back long before the classical Greek period to the pre-Socratics, the Egyptians, and the Babylonians, mathematics and religion have often been connected. These links exist not only in Western thought but all through the history of world civilization and in all religions and philosophies, including Chinese, Indian, Muslim, Jewish, Christian, and Buddhist traditions. Religious thinkers and philosophers seeking a conception of the "Absolute," the "Infinite," or the "Ultimate" have often believed that they found inspiration or a basis in mathematics.

However, religion is not the same as mysticism, which is usually defined as the belief that direct knowledge of reality or God comes through immediate insight or illumination, rather than through ordinary sense perception or rational analysis. Descartes was religious—he believed in God—but he was not a mystic; he thought he could prove the existence of God through rational thought, and actually attempted such a proof.[1] A true mystic would not be interested in such an effort.

Some people who mixed mysticism and mathematics were in such states of irrational intoxication that they may actually have been mentally deranged; others were geniuses. Sometimes the borderline between the two groups is not well defined. But the list of scientists and philosophers who are recognized in history for making genuine contributions to their fields but who also, at least at certain moments, expressed mystical yearnings is long and impressive; it includes Pythagoras, Blaise Pascal, Hermann Weyl, Arthur Stanley Eddington, Alexander Grothendieck, and many others. The astronomer and mathematician Eddington spoke of his "mystical contact with Nature," and of "the eye of the body or the eye of the soul."[2] Hermann Weyl commented in his *God and the Universe* that "mathematics . . . lifts the human mind into closer proximity with the divine than is attainable through any other medium."[3] Such observations understandably drew the censure of many of their scientific colleagues, or, more often, were quietly ignored.

It would be impossible to explore here all the varieties of links among religion, mysticism, and mathematics. There is an enormous literature on the subject, both good and bad.[4] Our concern in this book is the role of mysticism in the reception and development of set theory in Russia, and therefore we will concentrate on our Russian Trio.

The most important figure here is Luzin, one of the great mathematicians of the twentieth century. Reconstructing Luzin's philosophical development is not easy—he published almost nothing explicit about it and was sometimes secretive, especially during the Soviet period; before his death in 1950 he burned his diaries so that his innermost secrets would never be known. But when one studies his letters and mathematical papers (see the Appendix), many of which have been preserved in archives and family collections, a story of deep philosophical and mystical commitments emerges. Luzin's intellectual ties with his teacher Egorov and his onetime fellow student Florensky, both Name Worshippers, were important in constructing his worldview.

An early and revealing indicator of Luzin's interest in mysticism can be found in a letter that he wrote to his new wife, Nadezhda Mikhailovna, on June 29, 1908.[5] He had just read Florensky's thesis, written at the Moscow Theological Academy, entitled "On Religious Truth." Luzin was thrilled by the manuscript and wrote to Nadya with great excitement:

> In addition to discussing understanding through the senses ("physics," "natural science") and understanding through the mind ("mathematics," "logic") Florensky has given equal right to another kind of understanding, which you never hear about at the university, namely "intuitive-mystical understanding."

Here Luzin revealed his attraction to "other ways of knowing," a mark of the true mystic. He looked for books and writings that might help him toward that elusive goal.

Although giving "names" to sets and thereby endowing them with "reality," a concept taken from the Name Worshipping movement, was influential in Luzin's approach to mathematics, we see in this 1908 letter that he was drawn to mysticism before he had ever heard of Name Worshipping. The Name Worshippers merely increased his interest in mysticism; they did not create it. During the years 1908–1910 he studied simultaneously the deepest issues of set theory and the classics of religious mysticism.

On April 12, 1909, Luzin wrote Florensky that he was planning to study Plotinus (204–270 C.E.), a "mystic . . . who is no stranger to deep logical work required for a real worldview." Plotinus appealed to Luzin because he was on the borderline between the classical period of the ancient world and the beginning of Christian mysticism, so strong in the Middle Ages. As a mathematician, Luzin felt an affinity with the disciplined logic of a Greek philosopher and believed that he would not be as sympathetic with the overwhelmingly religious thought of the mystics of the medieval period.[6] He was obviously seeking a way to unite rigor with mysticism—not an easy task.

Plotinus is often called the founder of Neoplatonism. He based his thought on the writings of Plato but developed a complex view of the world that emphasized a spiritual element. Within his system were three central constituents: the One, the Intelligence, and the Soul. The One transcends all beings and, through a process which Plotinus called "emanation," gives existence to all beings, including Intelligence and the Soul. A part of Plotinus's worldview that must have been particularly striking to Luzin was that Intelligence, through contemplation of "The One," gives birth to forms *(eide)* which serve as the referential basis of all other existents. Thus Plotinus believed that the mind plays an active role in shaping the objects of its perception, rather than just being a recipient of sense experience. Luzin would later express the view that "the natural sequence of numbers does not have by itself an absolutely objective existence . . . it exists as a function of the mind of the mathematician." The similarity here with the thought of Plotinus is obvious.

On September 22, 1910, Luzin wrote Florensky that he had spent the entire previous summer reading William James's book *The Varieties of Religious Experience*, saying that he admired the work very much. One of the central chapters in James's book is on religious mysticism. Luzin had been so stimulated by his study of the particular form of mysticism expressed by Plotinus that he began searching for a more general analysis of mysticism, finding it in James. And Luzin would have immediately noticed that James, in his discussion of mysticism, referred to Plotinus, quoting him as saying: "In the vision of God what sees is not our reason, but something prior and superior to our reason."[7]

James described two specificities of mysticism that must have resonated with Luzin's proper mathematical interests. First, James was concerned with the capacity of humans to have direct access to knowledge, and he spoke of certain "states of insight into depths of truth unplumbed by the discursive intellect. They are illuminations, revelations . . . all inarticulate though they remain."[8] James observed that these insights have a "noetic quality," meaning that those mystics who possess them believe they are "states of knowledge." The word "noetic" comes from the Greek *nous* (νούς or νόος, a philosophical term for mind or intellect) and would have reminded Luzin of Plato's view of mathematics, where "one seems to dream of essence." Moreover, James characterized a mystical state as being "ineffable," that is, it "defies expression, that no adequate report of its content can be given in words."

Such ineffability could exist for two reasons, either because the state defies any possible description or because it defies any logical rule needed for verbal expression. The first case often arises in the heart of intense mathematical research, moments when mathematicians refer to "marvelous intuitions" without any possible explanation of their origins. The second case had a specific example in the antinomies which arose as soon as Cantor discovered the whole hierarchy of infinities; he then went on to prove the mathematical impossibility of a largest cardinal number, the "set of all sets"—a notion that was either incomprehensible or, at certain periods in Cantor's thinking, a fitting symbol for the "One" of Plotinus.

We do not know when Florensky and Luzin first learned of Name Worshipping, although it is clear that both were interested in the significance of "naming" as early as 1906, when they bestowed new names on each other as a sign of a brotherly bond.[9] Furthermore, in 1906–1907 Florensky wrote the manuscript of his book "Holy Renaming." As a seminarian studying for the priesthood, Florensky surely would have noticed the publication in 1907 of Ilarion's classic work on Name Worshipping, *On the Mountains of the Caucasus*, but at first the book was known primarily to a small circle of Orthodox monks. The Mt. Athos events which brought Name Worshipping to the attention of the reading public in Russia did not occur until 1913. However, we know that Florensky was interested in the movement before then because he and his friend M. A. Novosyolov analyzed the works of the Name Worshippers in October 1912.[10] Luzin, an intimate friend of Florensky in these years, would have known about these developments. Then in March 1913, Florensky began speaking out publicly in favor of the rebellious monks on Mt. Athos, several months before the invasion of the monastery by the marines. Florensky became one of the most convinced and influential believers in Name Worshipping, writing extensively on the subject.[11] At the monastery outside Moscow, he came in contact with participants in the Athos events.

In the specific form of mysticism represented by the Russian Name Worshippers, the links between mathematics and religion were carried to a new level. In the early twentieth century mathematicians were perplexed by the possibility of new kinds of infinities. Georg Cantor suggested these new infinities and made them seem real by assigning them different names. For some people the very act of naming these infinities seemed to create them. And here the Russian Name Worshippers had their opening: they believed they made God real by worshipping his name, and the mathematicians among them thought they made infinities real by similarly centering on their names.

Pavel Florensky communicated the ideas of the Name Worshippers to Luzin and Egorov and translated them into mathematical parlance. Florensky maintained that the Name Worshippers had raised the issue of "naming" to a new prominence in a way that had relevance for mathematics. To name something was to give birth to a new entity. Florensky was convinced that mathematics was a product of the free creativity of human beings and that it had a religious significance. Humans could exercise free will and put mathematics and philosophy in perspective. The famous statement of Georg Cantor that the "essence of mathematics resides in its freedom"[12] clearly had a strong appeal for Florensky. Mathematicians could create beings—sets—just by naming them. To take a simple example, defining the set of numbers such that their squares are less than 2, and naming it "A," and analogously the set of numbers such that their squares are larger than 2, and naming it "B," brought into existence the real number $\sqrt{2}$. Similar namings can create highly complex new sets of numbers (see p. 121 and the Appendix).

Florensky saw the development of set theory as a brilliant example of how naming can result in mathematical breakthroughs. A "set" was simply an entity named according to an arbitrary mental system, not an ontologically existing object. When a mathematician created a set by naming it, he was giving birth to a new mathematical being. The naming of sets was a mathematical act, just as, according to the Name Worshippers, the naming of God was a religious one—and the operation was performed in the same way. A new form of mathematics was being born, said Florensky, and it would rescue mankind from the materialistic, deterministic modes of analysis so common in the nineteenth century. And indeed, set theory, new insights into continuous and discontinuous phenomena, and discontinuous functions became hallmarks of the Moscow School of Mathematics.[13]

The idea that "naming" is an act of creation has a long history in religious and mythological thought. It has been claimed that the Egyptian god Ptah created by naming with his tongue that which he conceived in his head. In Genesis we are told that "God said, 'Let there be light'; and there was light." In other words, he named the thing before he created it. Names are words, and the first verse in the Gospel according to John states: "In the beginning was the Word, and the Word was with God, and the Word was God." In the Jewish mystical tradition of the Kabbalah (Book of Creation, Zohar) there is an emphasis on creation by naming, and the name of God is considered holy.[14]

Andrei Bely, student of the mathematicians Egorov and Bugaev (who was also his father) and the classmate of Luzin and Florensky, made the following statement on the power of naming: "When I name an object with a word, I thereby assert its existence." We can then ask, Does this apply both to mathematics and to poetry? If the object is a new type of infinity, does that infinity exist just after you name it?

A direct linguistic connection existed between the religious dissidents in Russia who emphasized the importance of the names of Jesus and God, and the new trends in Moscow mathematics. As we have seen, Luzin and Egorov were in close communication with French mathematicians who had similar concerns. In 1904 Henri Lebesgue introduced the concept of "named sets." He spoke of "naming a set" *(nommer un ensemble)*, and such a set he then called a "named set" *(ensemble nommé)*. The Russian equivalent was *imennoe mnozhestvo* ("named set"). Thus the Russian word *imia* ("name") is found in the terms for both the new type of sets and the religious practice of *imiaslavie* ("Name Worshipping"). And indeed, much of Luzin's work on set theory involved the study of effective or "named" sets.[15] To Florensky, this meant that religion and mathematics were moving in the same direction.

Luzin placed great emphasis on the significance of naming as he did his mathematical work, as is shown in more detail in the analysis of the material from his archive in the Appendix. One Western mathematician who studied Luzin's personal archives in Moscow observed that he

> frequently studied the concept of a 'nameable' object. . . . To Luzin the continuum conjecture was merely one aspect of the general problem of *naming*. . . . Luzin was trying very hard to *name* all the countable ordinals.[16]

At one point Luzin wrote in his notes, "Everything seems to be a daydream, playing with symbols, which however, yield great things." Elsewhere, he scribbled in infelicitous but understandable French, "nommer, c'est avoir individu" ("naming is to have individuality").[17]

Among French mathematicians Luzin's closest friend became Arnaud Denjoy, a man with whom he shared some of his religious interests in naming. When Denjoy asked Luzin to be the godfather to his son René, the Russian mathematician was delighted and replied:

> I heartily thank you for the honor and friendship you show me by choosing me as the godfather for your little René. . . . As you know well, christening has for me a profound meaning. The Universe cannot be reduced to social and physical forces, there remains a much more important part: the living soul.[18]

And then Luzin suggested that little René Denjoy be given an additional religious name, and proposed "a name common to the catholic and orthodox religions, such as 'Pierre.'"

Both the French and the Russian mathematicians were wrestling with the problems of what a mathematical object is, what mathematicians are allowed to do. Lebesgue wrote to Borel in 1905, "Is it possible to convince oneself of the existence of a mathematical being without defining it?" Florensky saw this question as the analogue of "Is it possible to convince oneself of the existence of God without defining him?" The answer for Florensky—and, later, for Egorov and Luzin—was that the act of naming in itself gave the object existence. Thus "naming" became the key to both religion and mathematics. The Name Worshippers gave existence to God by worshipping his name; the mathematicians gave existence to sets by naming them. This approach was particularly applicable to the transfinites of the Second Class, rejected by the French mathematicians, and to the complex hierarchy of new sets introduced by Luzin, Suslin, and their followers, starting with analytic sets. It is striking to see the confidence of the claims made by Luzin in his personal notes (see the Appendix):

> We, in our mind, consider natural numbers *objectively existing*.
>
> We, in our mind, consider the totality of all natural numbers *objectively existing*.
>
> We, finally, consider the totality of all transfinite numbers of the Second Class *objectively existing*.

We want the following: having assumed that we face [them], we connect with each of the transfinites a definition, a "Name" —and moreover uniformly for all those transfinites we are considering.

Other mathematicians have noticed the importance of naming objects and concepts in mathematics in completely different contexts, with no relationship to Name Worshipping. The eminent French mathematician Alexander Grothendieck (1928–), for example, also gave attention to the process of naming (see his *Récoltes et Semailles*).[19] One observer of Grothendieck's work wrote, "Grothendieck had a flair for choosing striking evocative names for new concepts; indeed, he saw the act of naming mathematical objects as an integral part of their discovery, as a way to grasp them even before they have been entirely understood."[20] Ineffable concepts that are sometimes linked to mystical inspiration and resist definition must be named before they can be brought under control and properly enter the mathematical world.

Thus, when we emphasize the importance of Name Worshipping to men like Luzin, Egorov, and Florensky, we are not claiming a unique or necessary relationship. We are simply saying that in the cases of these thinkers, a religious heresy being talked about at the time when creative work was being done in set theory played a role in their conceptions. It could have happened another way; but it did not.

6

The Legendary Lusitania

> And we, the wise men and the poets
> Custodians of truths and of secrets
> Will bear off our torches of knowledge
> To catacombs, caverns and deserts.
> —*L. A. Lyusternik, former member of Lusitania, describing his teachers Egorov and Luzin in the first years after the Bolshevik Revolution*

SHORTLY BEFORE World War I, Luzin and Egorov began to offer together an undergraduate mathematics seminar at Moscow University that was the embryo of what became known as the Moscow School of Mathematics. The circle of eager students that formed around them and continued through the early 1920s took on the name "Lusitania." The origin of the term is not clear, despite much discussion of the topic. Among later Moscow mathematicians, the most common explanation was that the student circle took its name from the British ocean liner *Lusitania* that was torpedoed by the German submarine U-20 on May 7, 1915. This event caused a great international outcry and was one factor influencing the entry, almost two years later, of the United States into World War I. The problem with this explanation is that, according to several participants in the seminar, the term "Lusitania" was used at the mathematics seminar *before* the sinking of the ocean liner.[1] Perhaps that event gave the name extra meaning.

Another explanation for the name, perhaps the most logical one, is that it involved a play on the word "Luzin," and was actually, in its first form, "Luzitania." However, this idea is also disputed, especially by those who note that Egorov was, at least in the early years, the senior professor of the seminar. Luzin was quoted by the students as saying that "Egorov is the chief of our society" and "our discoveries belong to Egorov."[2] It seems unlikely that Luzin would have agreed to the naming of the seminar after himself as long as Egorov was present, as he was throughout its history. Here again, though, the name gained extra strength through the resemblance of "Luzin" and "Lusitania," especially after Luzin emerged as the intellectual leader of the group.

Yet a third hypothesis, also disputed, is that the term goes back to the ancient province of the Roman Empire (after which the ship *Lusitania* itself was named) in what is present-day Portugal and Spain.[3] This province of Lusitania has a colorful history that might have appealed to the romantic young students at the time of the creation of the seminar, before the Russian Revolution. But all these explanations for the name "Lusitania" are speculations. We simply do not know why the seminar was named as it was.

A sense of the place of religion in the concerns of the Lusitanians (before the impositions on religion that followed the Soviet takeover) can be seen in early descriptions of the group.[4] According to one of them, the early Lusitanians acknowledged two leaders: "God-the-father" Egorov and "God-the-son" Luzin. Students in the society were given the monastic titles of "novices." Another historian wrote, "There was clearly a strong sense of belonging to an inner circle or secret order."[5] All the principals and novices went to Egorov's home, the apartment on Boris and Gleb Street, three times a year: on Easter, Christmas, and his name-day (underlining again the importance of names). The intense camaraderie among the Lusitanians was facilitated by Luzin, who was described as extroverted and theatrical, and who inspired real devotion among students and colleagues.

The Legendary Lusitania

Egorov, the senior member, on the other hand, was more reticent and formal.

For a while Egorov's and Luzin's chief assistants in managing Lusitania were three students, each with his own function: Pavel Alexandrov was the "Creator," Pavel Uryson the "Keeper," and Viacheslav Stepanov the "Herald" of the mysteries of Lusitania. All three of these students went on to become mathematicians of note; all three, along with their teachers Egorov and Luzin, would be included in authoritative listings of deceased scientists of world rank and in world scientific literature.[6]

Lusitania put Moscow on the mathematical map of the world. Before World War I there was only one mathematician at Moscow University whose name was well known to mathematicians in western Europe: Dmitri Egorov. By the end of the 1920s, there was a constellation of such mathematicians. And by 1930 Moscow had become one of the two or three most concentrated focal points of mathematical talent anywhere on the globe. Even many years later, in the 1970s, a leading Western mathematician observed that "Moscow probably contains more great mathematicians than any other city in the world," mentioning Paris as the only competitor, and noting that in several other countries which also have great mathematical strength, such as the United States, the mathematicians are more scattered geographically.[7]

A remarkable characteristic of Lusitania was the youth of the students who belonged to the group. When Lev Shnirel'man, who eventually made significant contributions to number theory and the calculus of variations, joined Lusitania he was just 15 years old. Andrei Kolmogorov, one of the great mathematicians of the twentieth century, was 17 or 18 when he first came to the attention of Egorov and Luzin. Other youths who joined Lusitania when they were 18 or younger and who later became notable mathematicians included Lazar Lyusternik, Pavel Uryson, and Pavel Alexandrov.

Many of the members of Lusitania, despite their already apparent mathematical talents, were thus adolescents, and their mathematical styles were still malleable; Luzin and Egorov shaped those styles.

Building of the old Moscow State University where the Lusitania seminars were held.

Youthful silliness and hilarity accompanied deep investigations into the foundations of mathematics. The students were so devoted to their teachers' studies of set theory that they made fun of mathematicians who worked in other areas, giving their topics comic titles such as "impartial differential equations," "theory of improbability," and "different finitenesses."[8] And as young people they were particularly susceptible to the demonstrative charms of their teachers, especially Luzin. Some of them were very young women. Nina Bari, the first woman ever to graduate from Moscow University (not from the special "Women's Courses" that existed before 1917) and later internationally known for her work on trigonometric series, was only 17 when she joined Lusitania. She and the other female students—I. A. Rozhanskaia, B. I. Pevzner, T. Iu. Aikenval'd—adored Luzin, and everyone knew that it was not just Luzin's mathematical abilities that attracted them. Bari's death over forty years later would be linked to that of Luzin.

The building that housed the department of mathematics of Moscow University in the first decades of the twentieth century, and the place where the Lusitania seminars were held, was built in a grand style in the 1830s on the order of Tsar Nicholas I. The imposing structure, with a view of the Kremlin, has two wings; one of them contained in tsarist times (and contains again today) the Church of St. Tatiana the Martyr, the university church, while the other wing houses the university library. Today, as one enters the front entrance of the main building one finds a grand marble staircase leading upward to a large, elegant room with a skylight that illuminates much of the building. On the second floor, where the Lusitania seminar was usually held, an expansive gallery with ornate pink stone columns surrounds the central staircase, and decorative frescoes adorn the arched ceilings. The floors are tile and provide room for a complete circumnavigation of the central building around the staircase, perfect for walking or, as we will see, skating. The second floor also has a large amphitheater which was named, in succession, the Big "Theological," "Communist," and "Academic" Auditorium, each name representing the ideology of the tsarist, Soviet, and post-Soviet times. This impressive and pleasant setting has been the scene of misery, shabbiness, and even ruin; there were shortages of food and clothing and inadequate maintenance in the twenties, political arrests in the twenties and thirties, and later, in October 1941, heavy damage from a German firebomb that penetrated the skylight. The interior of the building has now been restored to its earlier glory, and today the faculty of journalism occupies the space earlier taken by mathematics, which moved in the 1950s to the new university buildings on Sparrow Hills overlooking the city. But this old building is still regarded as the birthplace of the famed Moscow School of Mathematics.

In the seminar, Luzin was a showman who knew how to enthrall his class members. He would enter the room of expectant students, take off his coat, and speak to them in his academic gown. People often remarked that he had a "mystical view of the universe." He would make statements like "before our intellectual gaze there opens a vision of extraordinary beauty."[9] And then he would speak of transfinite numbers and of sets possessing subsets each of which was somehow equal to the whole. (For example, there are as many points in a segment of a line as there are in the larger line of which it is a segment.) One of his students observed, "Other professors show mathematics as a beautiful completed structure, and we can only admire it. Luzin shows it in its incomplete form, he awakens a desire to take part in its development."[10]

Luzin's approach to lectures was radically different from that of other Moscow University professors. Most of his colleagues simply read to the students, staring down at pages often yellowed with age, and hardly acknowledging the presence of an audience. (In the Russian language one does not usually speak of "giving a lecture," but rather of "reading" it, *chitaet lektsiiu*.) The boring character of most university lectures was so well known that some students never attended class, appointing a fellow student to take notes to be shared or obtaining written copies of the lectures from the university porter. However, if we are to believe the accounts left us by Luzin's former students, they were eager to attend his classes. And Luzin made sure that they felt involved. He would begin a proof at the blackboard, pause, and then say, "I cannot recall the proof; perhaps one of my colleagues could remind me." This was a challenge that the class felt obligated to meet. One student would jump up, go to the blackboard, attempt the proof, fail, and then sit down with a red face. Another would get up, perhaps a 17-year-old, successfully write the proof on the blackboard while the entire class stared enviously, and then sit down. Professor Luzin would turn to that student, bow slightly, and say "Thank you, my colleague." Luzin treated the students as intellectual equals, and his teaching led them to prepare for and anticipate coming lectures. One of them later asked, "Had Luzin [really] forgotten the proof, or was it a well-constructed game, a method of arousing activity and independence?"[11] They never knew.

Luzin overcame the traditional chasm between professors and students at Moscow University. When he finished a lecture, it often did not actually end. The students would surround him, asking questions

The Legendary Lusitania

Luzin's apartment (second floor) on Arbat street, Moscow.

and making suggestions; follow him down the large entry stairway; and then walk with him down Mokhovaia and Arbat streets to his apartment at the corner of Arbat and Afanas'evsky Boulevard. (The building still stands today and has a plaque on it saying: "The creator of the Moscow School of Mathematics, N. N. Luzin [1883–1950], an outstanding scientist, lived in this house from 1908 to 1925.") There Luzin's wife Nadezhda would be waiting with tea (and pastries, if she could buy the ingredients in those hard times), and the conversations would continue far into the night.

Luzin's approach was wonderful, and it energized a generation of Russian mathematicians. Lusitania was probably the most creative and fascinating chapter in the entire history of Russian mathematics (and there are other glorious chapters). One must admit, however, that it also had flaws, some of which would later become significant. Luzin was a very emotional man, and he tended to be either an enthusiast for another person or alienated from him or her; there was very little middle ground. And he was intensely proud of his role as the *maître*, the master of the group. If a student broke away and went in another direction, perhaps entering a field of mathematics in which Luzin was not interested, he felt hurt, maybe even betrayed. This attitude would be the source of later misery for him.

After the Communists came to power in November 1917, Egorov and Luzin dropped any explicit reference to religion in their discussions of mathematics with their students, but they retained the philosophy of mathematics that had been connected with it. Later mathematicians who were products of the Moscow School of Mathematics often did not share, and probably did not even know about, the religious impulses that were so important to Egorov and Luzin. However, some Western mathematicians who have attended seminars even recently at Moscow University have returned marveling over the intense, almost religious, aura that often permeates them. Perhaps something of the spirit of Lusitania is still alive.

Egorov was reserved and strict, but in a way that was the perfect foil to Luzin's approach. Egorov was a professor of the old style; he always dressed in a coat and tie and held himself aloof from student frivolity. But he too was devoted to his students and was more conscientious even than Luzin in attending to their actual practical needs: helping them to find jobs, finding fellowship opportunities in France and Germany, lobbying for stipends at the university. And several times a year he would invite the students to his apartment, also nearby—a much more formal occasion than the visits to Luzin's apartment. During these visits the students of the early Soviet period would catch a glimpse of the high culture of the pre-revolutionary Russian intelligentsia. Although Grzhimali, Egorov's father-in-law, and his Stradivarius violin were no longer around, Egorov's wife Aida or his wife's sister Natalya would play the piano for the students, and they might even sing university songs, now abandoned in Soviet times. They also might see the Bible and religious books on Egorov's desk, although no one would mention them.

One of the songs the students would sometimes sing at the Egorov apartment was "Gaudeamus igitur," which contained phrases that seemed subversive in the new Soviet environment:

The Legendary Lusitania

> Vivat Academia,
> Vivant professores!
> Vivat et respublica,
> Et qui illam regit!
> Vivat nostra civitas![12]

Lusitania was remarkable for many reasons, among them the fact that one of the great movements in world mathematics was created in the worst possible conditions. Russia was afflicted at this time by war, revolution, civil war, famine, and shortages of every kind. Although food was especially scarce, senior professors were granted special allotments, and Egorov used to give students portions of his professor's rations. Amazingly, Egorov kept several cats and a dog in his apartment, and they got part of his rations as well. Egorov was known to have digestive problems, and he evidently survived on something simpler. When the American Relief Association (ARA), directed by Herbert Hoover, heard that some of the students at Moscow University were starving, it opened a student cafeteria offering free food which served hundreds. But the Soviet government closed it down, insisting that the stories of hunger among students were false rumors.[13]

The classrooms where Luzin and Egorov lectured were often unheated. The rector (president) of the university, M. Novikov, issued an administrative ruling that if the temperature in a classroom fell below -5 degrees (23 degrees Fahrenheit), classes in that location were canceled.[14] Luzin's and Egorov's students ignored the ruling. They came to the seminars dressed in heavy sheepskin coats, if they had them, or in multiple layers of shirts and sweaters if they did not. If a white spot appeared on one of the student's faces during a lecture, others rushed to rub it with their hands to prevent frostbite. One of the students wrote a little poem about the situation in 1921:[15]

> Severe twenty-first year,
> Moscow University . . .
> Although I was then so young

> Although in a sheepskin coat I dressed,
> Yet... brr... What devilish cold...
> Here only the arguments are heated.
> With unquestioning faith I joined
> The young and noisy group.
> Despising classical analysis,
> Here they are carried away by the modern.
> Forward! Have confidence in yourself!
> The Lord himself—Professor Luzin—
> Shows us the pathway to research!
> The days of legendary Lusitania,
> Days of enthusiasm and striving...
> All of us infatuated with Luzin
> Jealous of each other on his account—
> Lest shine but a little
> Mathematical merit.
> I remember how each time—
> What emotion gripped one,
> Arriving at the appointed hour.

Outside the classroom on the second floor was a very wide corridor. Looking for a way both to keep warm and to get some exercise, the students carried snow up to the corridor, covered the floor with it, and then poured water on top. After a little smoothing with a broom of twigs and the quick onset of freezing, they had a skating rink in the mathematics department. Wearing old clamp-on skates, the students would await the arrival of their professors, singing while they glided on the ice around the central staircase under the skylight.

The problems they faced were not just those of cold and food shortages, but also political ones. Gradually the new Soviet government was imposing a new order on educational institutions. One evening in the spring of 1919, during a storm with lightning and rain, two trucks filled with communist workers pulled up to the university at midnight. The workers entered the Church of St. Tatiana the

The Legendary Lusitania

Interior of the Church of St. Tatiana the Martyr, Moscow.

Martyr, the university chapel, where they confiscated the crosses and icons of the church. They also attempted to remove the large inscription on the outside of the building, which read "The Light of Christ Illuminates Everyone." Accomplishing this required several hours, but eventually the work was done. The president of the university, Novikov, witnessed this event and later said that he saw "embarrassment" on the faces of the workers who had been assigned this task.[16] The church was converted into a student theater and social club and remained that way until 1995, when it was restored and the inscription was again placed on the building. For a short time in the post-Soviet period, as we described in the Introduction, the basement of the restored church became a sacred spot for the renewed Name Worshipping movement, with photographs of Egorov and Florensky displayed on the walls.

How do we explain the fact that an explosion of mathematical creativity of international significance occurred in conditions of political oppression, material deprivation, bitter cold, and even famine?

Something almost magical happened in Lusitania; it was like an alchemical reaction involving necessary ingredients—in this case, gifted professors and students—and also a dash of mysticism. A closer look at the situation of Egorov and Luzin, and how they reacted to it, will help to illuminate the phenomenon of Lusitania.

Although Egorov was older than Luzin, and was his teacher, both of them belonged to the pre-Revolutionary generation. Both were full professors at Moscow University when the Communists came to power. Therefore, in the eyes of the Communists they were members of the "bourgeois intelligentsia," at heart enemies of the new order. The fact that before the Revolution both were known to be deeply religious and were close friends of several priests, including Florensky, only deepened their difficulties.

In the first few years after the Revolution, Egorov and Luzin witnessed threatening political forces moving closer and closer to them and their beloved institution, Moscow University. Dozens of university teachers and students were arrested. Both the president of the university, M. Novikov, and their immediate superior, V. Stratonov, the dean of the physics-mathematics faculty, were seized and taken to the Lubianka Prison, rapidly becoming infamous as a political jail. (It was so close to the university that the secret police could force the prisoners to walk to the waiting cells.) One of their mathematics teaching colleagues, A. A. Volkov, was arrested and almost immediately shot, without even a pretense of a trial. Several of their colleagues committed suicide. The university chapel, important to them as religious believers, was looted and then converted to a nonreligious use. Many priests were imprisoned. The autonomy of the university, a hard-won recent achievement during the brief period of the previous Provisional Government, was eliminated.

What were Egorov and Luzin to do in this situation? If they protested, they too would soon be arrested. What they chose was a path available to them as specialists in abstract mathematics that was not possible for many of their other colleagues. Their fellow professors in experimental science could not pursue their specialties because they needed equipment and reagents that were now unobtainable.

The Legendary Lusitania

Their colleagues in fields like history and philosophy could not pursue their work because any conclusions or interpretations they could voice would almost certainly be judged ideologically unacceptable. But Egorov and Luzin needed no equipment and were engaged in highly abstract mathematics which the Communist authorities did not understand.

An illustration of the protection afforded by their abstract research is what happened to one of Luzin's and Egorov's students, Lev Shnirel'man. Egorov wrote a special letter to the minister of education, Anatoly Lunacharsky, urging a scholarship for Shnirel'man, who did not have enough money for food, and recommending him highly as a "specialist on Riemanian surfaces." Amazingly, Lunacharsky asked Shnirel'man to come to his office. There Lunacharsky said, "All right, I have examined your record, I see that it is outstanding, and I am going to give you this scholarship." Then, slapping his leg, Lunacharsky laughed and inquired, "Now, tell me, just what are these Riemanian surfaces, anyway?"[17]

Egorov and Luzin realized that if they buried themselves in their work, if they retreated into the ultimate ivory tower, they had a chance to do something worthwhile, both mathematically and culturally, even in the terrible conditions in which they found themselves. They could work with their young students to create wonderful mathematics, while at the same time representing to those students the best values of the pre-Revolutionary intelligentsia. They could no longer talk to the students directly about their religious commitments, but they could introduce them to what they saw as the inherently mystical and spiritual significance of mathematics. They possessed "secrets" which could not be fully divulged, but which could be taught by example.

Forty-five years later one of the former students of Egorov and Luzin, L. A. Lyusternik, by then a famous topologist, looked back on the years of Lusitania and tried to divine the feelings and motivations of his professors.[18] When he wrote about the topic it was still the Soviet period, and Lyusternik could not speak absolutely freely. But the message comes through clearly enough:

> These men had long since become set in their habits, tastes and ideals: it was not expected that many of them would welcome the new state of affairs right away. . . . As leaders of the bourgeois intelligentsia in the face of the imminent socialist revolution they felt that cultural values "that were known to us alone" were threatened with destruction.
>
> And we, the wise men and the poets
> Custodians of truths and of secrets
> Will bear off our torches of knowledge
> To catacombs, caverns and deserts.[19]

Lyusternik continued:

> To some old professors the cold and gloomy lecture rooms perhaps seemed to be indeed "catacombs and caverns." The phrase "withdrawing into an ivory tower" was in vogue at that time. It signified a flight from life's threats, isolating oneself in a secluded world of science or art. There were various forms of this withdrawal—the "insular attitude"—the desire to regard the university, department, institute and so on as an "island" in which an accustomed microclimate is maintained. There is no difficulty in finding evidence of this attitude at the university during the first years of Soviet rule, but important work was nevertheless done at that time.

Egorov and Luzin were not actually old professors at the height of Lusitania (in 1923, they were 54 and 40 respectively). But they were nonetheless products of the old professoriate, and their values diverged sharply from those of the new regime. In their lectures they could not explicitly say they believed that religion and mathematics were inextricably linked, but they could speak of the "mystical beauty" of the mathematical universe, of the ability of humans to create mathematical entities and concepts simply by identifying them and naming them; they described mathematics not as a codified body of truths related to the material world but as a product of human

The Legendary Lusitania

minds that was constantly being expanded. They invited their students to join them in this expansion, and they pointed to set theory as the most fertile field for this creativity. They encouraged the students to give new names to everything around them, and they did: they renamed themselves, their teachers, the institutions in which they worked, and the functions and sets they were learning. All of this was, of course, an expression of philosophical Platonism and idealism, opposed to Marxist materialism and realism—something that a few Marxist mathematicians would later notice and condemn; but for a while it had magical effects on the young students of Lusitania.

The students brought their own important contributions to Lusitania: energy, talent, and even cheerfulness in the face of what their teachers often saw only as gloom. Some of them hoped that the Soviet Union would indeed create the unprecedented great civilization and culture that the Bolshevik leaders promised. Nikolai Bukharin, the "favorite" of the Party, declared: "It is not only a new economic system that has been born. A new culture has been born. A new science has been born." When the president of Moscow University, M. Novikov, once went to the office of the Ministry of Education to protest the arrest of some of his professors, he asked, "Why is it necessary for you to be so destructive?" The official answered, "As a biologist, you must know that the birth of a human being is a bloody affair. So is the birth of a new political order." Some of the students were willing to accept this explanation.

In his memoirs written when he was an old man, Lazar Lyusternik remembered those days and asked, "Why were we so cheerful?" He then answered his own question: "Most of us were just children [Lyusternik joined Lusitania at the age of 17], but we found ourselves at the source of a great river of Soviet mathematics. We found ourselves in some sense involved in its inception. Certainly we were not then clearly aware of it, but we sensed something of the sort."

Thus, the mathematical and cultural sophistication of the teachers of Lusitania combined with and reacted to the youthful vigor of their students to produce a singular event: the birth of the Moscow School of Mathematics. Neither the professors nor the students alone could have approached this remarkable achievement. Two worlds met and produced something totally new.

All students on entering Lusitania were assigned a name taken from set theory.[20] Recruits were called $\aleph_0$. Each time a student had some sort of success, such as first publication, first lecture delivered to the Mathematics Society, graduation from the university, or passing the master's examination, that person's $\aleph$ number (aleph number) would be raised. Alexandrov and Uryson soon achieved the high rank of $\aleph_5$. Luzin himself was given the name $\aleph_{17}$. Egorov was $\aleph_\omega$, the "omega" subscript indicating that his status was higher than Luzin's but still not as high as the status of the Continuum. Papers that circulated among the members of Lusitania—what would now be called "pre-prints"—were often ornamented with the coat of arms of the author, an elaborate rendition of their $\aleph$ number.

Lusitania also had its musical anthem, something called the "Lusitania March." It was long thought that the composer was Nina Bari, but she denied authorship, claiming that S. A. Bernstein, later a professor of applied mathematics, was responsible. We do not know the entire refrain, but part of it was:

> Our deity Lebesgue,
> The integral our idol,
> Through rain and storm and snow
> We wend our merry way.[21]

The Mystery of "A": The Birth of Descriptive Set Theory

The first great creative moment of Lusitania came about with the birth of Descriptive Set Theory, an event that fully demonstrated the ability of the young Russian mathematicians in Moscow. For the few dozen people studying set theory at the beginning of the twentieth century the main problem, as Hilbert pointed out in Paris in 1900, was the Continuum Hypothesis (CH). A reasonable strategy to attack the problem (some other strategies would appear later)[22] was one that Cantor already had in mind around 1880: try to imagine all possible subsets of the Continuum (**R**, the real line) and give some kind of mathematical description of them. Accomplishing this seemed an enormous task because subsets of the line could be so different. For any such subsets Cantor needed to prove that one of two possibilities existed: either such a subset is denumerable, or it is in one-to-one correspondence with the points in **R**. Cantor succeeded in 1879 in proving the validity of this choice for any closed subset (as an application of a theorem he proved with Bendixson), but the general question remained open. Luzin himself confronted the problem and tried hard to give a description of the other subsets of the continuum, to *name* them. This work led to a new mathematical field based on employing transfinite numbers to *describe* highly complex subsets of the continuum—*Descriptive* Set Theory.

Borel had introduced in 1898 a very general family of subsets, the B-sets (later called Borelian sets). It was quite natural to ask if B-sets satisfied the Continuum Hypothesis. This problem was solved separately in 1915 by Pavel Alexandrov and by Felix Hausdorff. Alexandrov, a very talented Russian mathematician, was one of the early Lusitanians; he will be discussed in more detail in Chapter 8.

Hausdorff was born in Breslau in 1868 and would die in Bonn in 1942. In addition to being a brilliant mathematician, Hausdorff had a career in literature, writing under the pseudonym of Paul Mongré ("My Desire"). His double active interest in science and literature was not uncommon in German and Austrian culture of the time (called "Cacania" by the novelist Robert Musil). In 1914 Hausdorff published the first systematic treatment of set theory, *Gründzuge der Mengenlehre* (*Principles of Set Theory*), which had considerable influence during most of the twentieth century and established Hausdorff as an international authority in the field.

Hausdorff, working in Leipzig, and Alexandrov, working in Moscow under the guidance of Egorov and Luzin, proved the same result in complete independence (the front line of World War I was between them). Their achievement quickly became big news, since they had obtained the first result on the Continuum Hypothesis since Cantor and Bendixson.

After Alexandrov proved his theorem on B-sets, he gave a talk on October 13, 1915, at a student seminar at Moscow University attended not only by Egorov and Luzin but also by young researchers such as Pavel Uryson, Mikhail Suslin, and the Polish mathematician Waclaw Sierpinski (later very distinguished in the field). Sierpinski had been imprisoned in the war because he held an Austrian passport at the time, but he was rescued by Luzin and Egorov, who managed to get him transferred to Moscow to participate in their mathematical work.

Luzin then asked his student Alexandrov to prove the converse of his result: is any set constructed in this way a B-set, and is it possible to get all B-sets in this way? Lebesgue had suggested that all subsets of the line were probably of the B-type, but confirming this statement or giving a counter-example was very difficult. Alexandrov spent many months working on this problem, with no success, and grew irritated with Luzin for giving him an assignment of such complexity. This irritation was the seed of later difficulties between the two.

But then Mikhail Suslin, a young Lusitanian from Saratov with a strong character, took up the problem. Luzin gave him Lebesgue's seminal paper of 1905 to read. The first infant cry of the newly born Descriptive Set Theory came when Suslin rushed into Luzin's office with Lebesgue's paper in his hands. If one can imagine a hypothetical birth certificate for the new theory, it might look like this:

DESCRIPTIVE SET THEORY
CERTIFICATE OF BIRTH
Date: afternoon, October 1916
Location: Mathematics Department, Moscow University, Nikolai Luzin's office
Parents: Nikolai Luzin / Mikhail Suslin
In the presence of: Waclaw Sierpinski

Suslin excitedly told Luzin that he had found a mistake in Lebesgue's paper. At first glance, such an error seemed improbable. After all, Lebesgue was a great mathematician, intensely admired by both Luzin and Egorov, who had met him in Paris and attended his seminars. But Luzin also knew that Lebesgue had a very rich imagination, and occasionally he let it take precedence over his reason. Therefore, Luzin paid attention to what Suslin was saying. As Sierpinksi, who witnessed the event, reported, "M. Lusin treated very seriously this young student who was claiming that he had found a mistake in the paper of such an eminent scientist."[23] Together the two men, Luzin and Suslin, professor and student, studied the article closely. Yes, there *was* a mistake. It was Luzin's genius to see that this mistake might have important consequences.

Lebesgue's mistake revealed that since the projection of the B-sets of Borel might happen not to be a B-set, there was a chance that a new type of set would appear. And this is exactly what happened. Suslin, now working with Luzin, gave a precise example of a set not of the B-type; he named the new family of sets "A-sets," distinguishing them in that way from Borel's "B-sets." Later, Suslin and Luzin, using nondenumerable cardinals, created a whole hierarchy of subsets of the continuum.[24] It was as if sets, of kinds not known before, were emerging from a secret cavern, needing new names and notations.

Suslin, with the help of Luzin, defined in a clear way an operation that was radically new because in its definition for the first time one was using the set of all sequences of all integers; starting from, say, a family X of closed intervals, this operation, which he called "A-operation," was creating a new set called A(X) using the symbolism of trees to represent unions and intersections of sets.

Suslin was young, with a brilliant but naive mind, and he failed to see the danger in naming these sets "A-sets." Later many authors called the new sets "Suslin sets." But by using the initial "A," Suslin created a big problem because Alexandrov was able to say later that the "A" stood for his name, and that actually *he* should be given credit for finding A-sets. Suslin's death from typhus just three years later, in 1919, eliminated him as an active contender in the priority dispute that erupted. In Alexandrov's autobiography, published in 1979, he created a deliberate confusion between his proof of the theorem on

Nikolai Luzin (seated), Waclaw Sierpinski (standing, on left), and Dmitri Egorov (standing, on right), in Egorov's apartment on Boris and Gleb Street, Moscow.

B-sets in 1915 and the A-construction of Suslin and Luzin in 1916. Thus, in effect, he was claiming ownership of everything. Alexandrov was a significant mathematician in his own right, with many achievements that he could justly claim; it reveals a serious flaw in his character that he also felt compelled to claim the work of others.[25]

Only with the publication in 1999 of documents which Alexandrov thought would never be revealed did we learn that Alexandrov accepted at Luzin's trial that it was Suslin and Luzin who had defined analytic sets and had constructed the whole A-operation.[26] Thus, it was Luzin and Suslin who were the real parents of Descriptive Set Theory, with the French mathematicians Baire, Lebesgue, and Borel as grandparents.

The Legendary Lusitania

After the discovery of analytic sets in 1917, Luzin noticed that taking complements of a set X (a set of points which do not belong to X) posed interesting and difficult questions when X was an analytic set. He defined in 1925 a new class of sets, projective sets (which he strangely attributed to Lebesgue in the beginning); these were obtained by taking several projections and complements (in any order) of Borel sets. In this way he created a new level of the hierarchy, a huge family of new sets which emerged from the mysterious grotto of Lusitania's mathematics. In a series of "Notes" submitted to the Académie des Sciences in Paris, Luzin listed questions concerning these projective sets. With extraordinary insight, he asserted that Cantor's set theory was insufficient for solving some of them, fifteen or twenty years before such incompleteness would be established by Gödel and later Cohen. This intuition of Luzin was later criticized by Alexandrov, who disliked any restriction on mathematics and, in particular, on his rising domain of topology, for which he predicted a glorious, triumphant future (similar to the way fervent believers in the USSR saw the future of Soviet socialism itself). Alexandrov wrote a letter to Hausdorff in 1925 in which he criticized what he saw as the authoritarian and pessimistic views of his former master, Luzin.[27]

The Full Emergence of Lusitania in Russian Mathematics

In the spring of 1921 the Academy of Sciences in Petrograd (St. Petersburg, Leningrad) invited Moscow University and the Moscow Mathematical Society to participate in a conference celebrating the centenary of the birth of Pafnuty Chebyshev, one of the great figures in nineteenth-century Russian mathematics. The young Lusitanians were eager to accept the invitation and go to Petrograd, especially since they wanted to demonstrate the new strength of Moscow mathematics to the older and skeptical Petersburg school, represented by A. A. Markov, who was still alive. (Recall that Markov was the atheist enemy of scholars who attempted to link religion to mathematics, like his Moscow colleague P. A. Nekrasov.)

This invitation came at a difficult time in the history of the young

Soviet state. The civil war was not over, although the victory of the Reds over the Whites now seemed probable. Famine and disease afflicted many parts of the country. The mathematics students in Moscow had almost no money. Yet they managed to arrange the trip in a remarkable way, demonstrating that in the emerging Soviet system access and influence were more important than money. The students decided to try to convince the Soviet authorities to give them, free, an exclusive railroad car that would take them to Petrograd. They knew that the mathematician who had the most influence in the new Soviet government was Otto Shmidt, a communist, who held high positions in the Ministry of Finance, the Ministry of Education, and the Ministry of Food Supply. Furthermore, he knew Lenin personally. Perhaps Shmidt could help Lusitania go to Petrograd.

The Lusitanians invited Shmidt to come to the university to talk to them, which he did, being curious about this group of young Moscow mathematicians. His deep interest in mathematics is evident in his personal archive in Moscow today, which contains letters from many outstanding Russian mathematicians, including Luzin, Pavel Alexandrov, and Andrei Kolmogorov.[28] Shmidt was a colorful man, tall with a large beard, and he usually wore a leather jacket, as many early Bolsheviks did. In later years, in the 1930s, he became a famous polar explorer. He was also a notorious womanizer. It was once said of him that he became, on the same day, father to two children by different mothers in different cities. He arrived in 1921 to meet the students and their professors at the university with a new lady on his arm. When Luzin saw the enamored couple, he was amazed, and remarked that he had thought that a person like Shmidt who entered public affairs "lost the last epsilon of private life." The others told him that Shmidt had "such an epsilon on his arm every day."[29]

Shmidt loved the idea of the young Moscow mathematicians traveling to Petrograd in their own special railway car, and ordered that it be done. To a Marxist like Shmidt, this dispensation was a symbol of the privileges of the pre-revolutionary rich passing to the new generation of science-oriented Soviet citizens. An elaborate certificate was drawn up, with the seals of Moscow University, the Moscow

Otto Shmidt.

Mathematical Society, and Lusitania attached, and signed by Shmidt, to be shown to railway officials or police who might meet the group and question the authority of students to take over a railway car.

    The students were overjoyed, and they persuaded Egorov, Luzin, and their wives to come along. The students immediately began the process of "naming" everything that came their way. One of the students was dubbed "Commandant of the Carriage" and given the task of assigning compartments to the passengers. The compartments were also given names, such as "Pegasus' Stable" for the bachelor students and "Enclosure of the Professoresses" for the women. The students renamed the Academy of Sciences in Petrograd "the Lazaretto" (after Academician P. P. Lazarev, their host). When they arrived in Petrograd they found the beautiful city deserted, with no traffic in the streets, and grass actually growing between the cobblestones. A goat was grazing on Khalturina Street. They walked arm-in-arm down the middle of the city's most famous avenue, Nevsky Prospekt, the entire length from the Moscow train station to the Winter Palace on the Neva River, meeting only one vehicle during their stroll. Arriving at the famous Palace Square, which was (and still is) dominated by the enormous Alexander's Column, topped by an angel carrying a cross, they renamed the column "Lesser Men'shov" because they considered it less impressive than "Greater Men'shov," the tallest member of their group—the young professor Dmitri Men'shov, Luzin's first student. (A play on words was involved here, since *men'she* means "less" in Russian.) It was the time of the "White Nights" (late June, when the sun barely sets) in the city, and some of the students continued to walk the streets all night, splashing in puddles with bare feet when the rain came down. When Pavel Uryson rolled up his trousers under his long overcoat, his friends told passersby who saw only the overcoat that he had nothing on underneath.[30]

The Muscovites gave three lectures, one of them by Luzin, in which he emphasized the deep philosophical problems in mathematics, implying that creative work in the field could not be separated from profound questions of understanding. He deliberately tried to distinguish the new Moscow school from the old St. Petersburg one by the depth of its intellectual inquiry into the foundations of mathematics. Luzin's students were thrilled, and Uryson enthusiastically exclaimed to the others that "his" Moscow professor had "floored" the Petrograd mathematicians. The latter did not agree with this description, but it was clear to them that a new mathematical force was emerging in Moscow, just as that city had displaced Petrograd as the capital of the new Soviet state.

7

Fates of the Russian Trio

"*Who* is that?"

—Leon Trotsky, on seeing Father Pavel Florensky presenting a scientific paper at a Soviet conference while wearing his clerical robe

AFTER THE Communist revolution and on into the early 1920s, both Egorov and Luzin continued their teaching. For a while Luzin taught at the Polytechnic Institute in Ivanovo-Voznesensk, outside Moscow, but then returned to the city. Egorov never left Moscow, where he continued to nurture and instruct his students in the new trends in his field. Mathematics in Russia, and especially at Moscow University, prospered. However, both Egorov and Luzin were disturbed by the political events surrounding them. The secret police were working to eliminate all parties except the Communist Party, and the Soviet government was conducting a war on religion which seriously depressed Egorov and Luzin, both religiously devout. In 1922–23 the police hunted down and executed many priests. The leader of the Party and government, Vladimir Lenin, personally signed the orders for some of those executions. Egorov and Luzin were particularly worried about their close friend Father Florensky, still living outside Moscow near one of the most famous monasteries in Russia in the town of Sergiev Posad.

{ 125 }

Egorov, Luzin, and Florensky reacted differently to these repressions. Egorov and Florensky became tougher and more resistant in the face of political pressure; they continued to defend religion, even publicly, and pursued their meetings in the Name Worshipper circle. Florensky was the most defiant, refusing to take off his priest's robe, which caused the Soviet leader Trotsky to inquire at a meeting they both attended, "*Who* is that?" Egorov also continued his religious practices; he worked closely with Florensky in inspiring the "True Church" movement, which aimed at a religious revival in Russia despite the Soviet efforts to suppress religion. Luzin was much more cautious: he stayed away from the meetings of the Name Worshippers and tried to conceal his religious convictions, although he continued to be a fervent believer. He simply became less public about it. His relations with Egorov cooled, and he even distanced himself from his old friend Florensky, seeing the danger in being too closely associated with a priest.

Egorov, in contrast, was either extremely brave or perhaps insufficiently aware of how dangerous some of his behavior was in an increasingly militant Soviet Union. He criticized the action of the university administration in 1919 in closing the university church, the Church of St. Tatiana the Martyr, which had been a part of the old university campus since 1837.[1] And when the church building was converted into a student club, dance hall, and auditorium, Egorov pointedly refused to attend any of the events held there, considering them a desecration. Of course his stance was noticed, both by students and by university administrators.

Another anecdote serves to illustrate Egorov's antipathy toward the new Soviet authorities. A mathematics student named Vladimir Nikolaevich Molodshii once stopped Egorov in a hallway at the university and asked for help with a mathematics problem. Egorov immediately agreed, but as he was explaining the problem to Molodshii he noticed that the student was wearing on his lapel a pin saying "Young Communist League member." Egorov's facial expression suddenly changed; he interrupted his mathematics explanation, saying he was "very busy," and broke off the conversation. The offended

Molodshii went on to become a Marxist philosopher in the Academy of Sciences and a bitter opponent of the approach to mathematics represented by Egorov and Luzin, giving lectures and writing articles and books that sharply criticized them.[2]

In the 1920s, especially after Lenin's incapacity resulting from strokes in 1922 and his subsequent death in 1924, the intensity with which the Soviet regime would actively suppress religion was still somewhat in question, even among top leaders. The repressions occurred sporadically; periods of ferocity would alternate with times of grudging tolerance. A few religious institutions, such as the Trinity Monastery of St. Sergey in Sergiev Posad, with which Florensky was closely associated, managed to continue their existence. All Soviet leaders were opposed to religion, but some thought the best way to overcome it was through anti-religious education among the youth, not by the imprisonment or execution of religious believers. During the period of the New Economic Policy from 1921 to 1928, "capitalistic" elements (like small businesses) were tolerated in the economy even though the Soviet state was ideologically committed to state ownership and control of the entire economy; some people thought that a similar truce with religious elements was also possible, at least temporarily. The top leaders of the Soviet Union who were most open to such a possibility were Nikolai Bukharin and Anatoly Lunacharsky. Bukharin, the architect of the New Economic Policy, was a strong atheist who frequently pointed to the past mistakes of organized religion in the debates surrounding Copernicus, Galileo, and Darwin. Nevertheless, he appreciated the role that religion had played in the development of European culture. Bukharin hoped to overcome religion through persuasion, not coercion.

Anatoly Lunacharsky, minister of education in the Soviet government from 1917 to 1929, held similar views. He was very aware of the role of religion in European art and music, two of his favorite subjects, and he admitted his admiration of some aspects of "bourgeois" culture. He once remarked, "I am inclined to think that Marxism as a philosophy is the new and last religious system—deeply critical and purifying and at the same time synthetic."[3] He had even

"*A Temple of the Machine-Worshippers.*" Drawing by the constructivist artist Vladimir Krinski, c. 1925.

once, perhaps in a rash moment, supported the "God-Builder" movement, led by Marxist intellectuals who thought that the masses needed a psychological substitute for religion. It was too much to hope, the God-Builders suggested, that the peasant masses of early Soviet Russia would simply give up their religious yearnings. Instead, the God-Builders proposed Communism as an object of worship in place of God, a proposal that Lenin regarded as obscurantist nonsense. In their enthusiasm for making Communism an object of religious veneration, the God-Builders constructed altars for the adoration of Marxism and even proposed converting orthodox cathedrals into shrines to Communism and industrial machines. The noted constructivist artist Vladimir Krinsky illustrated such an attempt in a drawing made around 1925.[4]

There were only a few mathematicians in Russia in the early 1920s who were also Marxists. Two of the more prominent ones were Otto

Shmidt (1891–1956) and Ernst Kol'man (1892–1979), similar men in their political orientations but quite different in their tolerance for people of opposing views, and also different in their mathematical abilities (Shmidt was far superior to Kol'man). Shmidt, a member of the Communist Party since 1918, held several important government posts and also became director of the State Publishing House, editor of the *Great Soviet Encyclopedia* (together with Bukharin), and a member of the Central Executive Committee of the USSR. Shmidt spoke proudly of the importance of Marxist philosophy, even in mathematics, and lectured on the theme, but he also tolerated mathematicians of different views. He supported Luzin, for example, when the latter applied to the Rockefeller Foundation in the United States for a fellowship to do research on set theory in Paris. He also, as we have seen, arranged in 1921 the special railroad car for the Lusitanians.

Kol'man, on the other hand, was a militant Marxist of the new emerging Stalinist type, a man who would use extreme means to get rid of people he saw as ideological enemies. He was born and grew up in Czechoslovakia, where he was educated in mathematics at Charles University, but ended up in Soviet Russia when he was stranded there as a soldier at the end of World War I. He was an ideologue of a particularly dangerous type, a man who took his Marxism very seriously and considered all other philosophical viewpoints as threats to the Soviet state. He would play a sinister role in many events in Soviet history, and was a major accuser of Egorov, Florensky, and Luzin. At the same time, he had genuine intellectual interests, spoke and read four or five languages, and wrote several books on the history of science and mathematics that still deserve attention. No wonder he was sometimes referred to as "the dark angel." For a while, after World War II, Kol'man spent time as a prisoner in a Stalinist labor camp because he tried to interpret Marxism in his own way, rather than following what the Party leaders dictated. To the surprise of many people who knew of his dogmatism, in the late 1950s he emerged as a defender of cybernetics against ideological criticism.

Ernst Kol'man.

Eventually, still militantly following his own star, Kol'man became deeply disillusioned with the Soviet Union and emigrated to Sweden. Before he died in 1979 Loren Graham interviewed him several times, both in the Soviet Union and later in the United States. By that time Kol'man was wrinkled and elderly, and, as before, more interested in asserting his own views than listening to those of others. Just after his death, in 1982, a book of Kol'man's was published entitled *We Should Not Have Lived That Way*, in which he partially confessed to his crimes. He did not tell the story of what he did to Luzin and Egorov, but he did say, "In my time I evaluated many things, including the most important facts, extremely incorrectly. Sincerely deluded, I was nourished by illusions which later deceived me, but at that time I struggled for their realization, sacrificing everyone."[5]

In the 1920s and early 1930s, as time went on, the militant Marxism represented by Kol'man began to win out over the more moderate form advocated by Bukharin and Lunacharsky. But in the changing times Kol'man and others like him had to look for opportune moments to win battles over their adversaries. Kol'man always considered Egorov an ideological enemy but had been hesitant to attack him at Moscow University, where the professor had many supporters. Egorov was, after all, president of the Moscow Mathematical Society, director of the Institute of Mathematics and Mechanics at Moscow University, and one of the most famous mathematicians in the Soviet Union. In the early 1920s, however, Egorov took a step that gave Kol'man the opportunity he had been looking for. Needing money beyond his modest salary at Moscow University, Egorov began teaching part-time at the Civil Engineering Institute in Moscow. Now Kol'man saw his chance: knowing that Egorov was more vulnerable at the Institute than at prestigious Moscow University, he chose this place to make his first attack.

In 1924 Kol'man spoke at the Party meeting in that institute—where he knew he would have supporters—and described Egorov as a "reactionary supporter of religious beliefs, a dangerous influence on students, and a person who mixes mathematics and mysticism."[6] Because he was not a member of the Communist Party, Egorov was not at the meeting and did not have a chance to defend himself. When he was informed of the attack, he honestly and perhaps naively admitted that he was religious. He even defended his position, saying that educational institutions should tolerate people of diverse personal beliefs. As a result, he was dismissed from the Civil Engineering Institute.

Looking around for someone to replace Egorov in teaching mathematics at the Institute, its director identified Nikolai Chebotaryov as a likely candidate. Chebotaryov was young, only 30 years old, and did not appear to have any embarrassing ideological characteristics. He seemed to be, and apparently was, a loyal Soviet citizen. He had served as a "lecturer," evidently voluntarily, in the famous Chapaev division of the Red Army during the Civil War. He was not religious, and he defended secular views typical of members of the Russian intelligentsia. His young wife, Maria Smirnitskaia, was a student at a medical institute in Moscow, and both of them seemed to be representatives of the new generation of Soviet intellectuals. Their marriage had taken place in a civil ceremony, as favored by the Soviet regime, not in a church ceremony as desired by the groom's mother —a fact that contributed to permanent cool relations between Maria and her mother-in-law. The director of the Civil Engineering Institute was aware of these modern views of the young mathematician and his wife, and he offered Chebotaryov the teaching position earlier occupied by Egorov.

Chebotaryov, however, was a man of strong moral principles. Within a few weeks he learned who his predecessor had been. Although he had not studied with Egorov (he had been educated in Kiev under an outstanding mathematician, Dmitri Grave), and did not know him well, he was certainly aware of who Egorov was—one of the finest mathematicians in Russia. Chebotaryov inquired why such a talented and famous mathematician had been fired. He considered Egorov better qualified than he, and he was troubled by the whole episode. Several of Chebotaryov's acquaintances at the Civil Engineering Institute told him that Egorov had been attacked by Kol'man for being a religious believer, and was fired for that reason.

Chebotaryov discussed the situation with his young wife. Although both of them were unsympathetic toward religion, they were also idealistic young intellectuals; they agreed that it was unethical for Chebotaryov to take a position earlier occupied by such a qualified mathematician, and moreover one who had been fired for personal religious beliefs. In a remarkable demonstration of self-critical honesty, Chebotaryov submitted his resignation, saying that otherwise he could not live with himself.

The next few years were difficult for Chebotaryov and his wife, who had not yet finished medical school in Moscow. After unsuccessfully looking for work in Moscow, Chebotaryov finally found a job in distant Odessa, where he taught for a while. Then he received a good offer at the University of Kazan, an institution with a fine scientific tradition, but located far from Moscow on the Volga River. After his

Nikolai Chebotaryov.

wife completed her medical studies, she joined him in Kazan and took a position as a physician at a local Kazan hospital. Neither of them dreamed that they would meet Egorov again, but they did, under very dramatic and tragic circumstances.

During the 1920s the political atmosphere surrounding the Name Worshippers became more and more complicated. They continued to hold services in secret, including the practice of the Jesus Prayer, in many different parts of Russia. But now that the established church was under siege from the authorities, many local churches with no attachment to Name Worshipping also began to have secret services, calling themselves the "Catacomb Church" in honor of their religious predecessors like St. Tatiana, who hid from the pagan Roman emperors in the first centuries of the Christian era. As a result, members of the official church and members of the Name Worshipping heresy met more and more often in their hidden locations, drawn to each other by the fact that they were both trying to elude suppression. Under this threat, the theological differences over Name Worshipping seemed of only minor importance. The main goal was to keep religion alive. The Communist authorities made no distinction between orthodox believers and heretics; all of them were religious and therefore were to be eliminated if possible.

Then in 1927, something happened that changed the situation. Metropolitan Sergius, a leading church official, hoping to ensure the survival of the church, made a peace of sorts with the Soviet authorities and promised in the future not to oppose them in any way. This "compromise" was censured by many Orthodox believers, who saw it as an agreement with the Anti-Christ. A number of them continued to operate in secret as the Catacomb Church, refusing to recognize the agreement made by Sergius. The Name Worshippers became involved in this political and religious struggle because they remained, as before, in hiding. A fusion between some Name Worshippers and the Catacomb Church occurred in which their mutual opposition to both the established Church and the Soviet authorities was more important than doctrinal differences over the divinity of the names of God and Christ. As time went on, many Name Worshippers began to describe themselves in the same way in which the members of the previously orthodox Catacomb Church did: as the "Pure Russian Orthodox Church." One result of this merger was to make the Name Worshippers more conservative than before. Earlier, they had identified themselves primarily by their attitude toward the names of God and Christ; now, they were increasingly seen as conservative on many other doctrinal issues as well, including the family, marriage, sex, and religious liturgies. Egorov, however, remained committed to the original philosophical tenet of Name Worshipping.

In 1929 and 1930, with Stalin now firmly in control of the government, the Soviet authorities again made mass arrests of religious believers, and this time the sweep included the Name Worshippers in and near Moscow. Kol'man once again went on the attack against Egorov, and now he assaulted him in his main position of strength:

Fates of the Russian Trio

Moscow University. Kol'man mixed his criticisms of Egorov's and Luzin's religious and philosophical positions with more overtly political accusations. Egorov was, he said, a "saboteur" or "wrecker" *(vreditel')*. This term was being used with great effect in other legal prosecutions, such as that of the "industrial wreckers" in the "Shakhty Trial" of 1928. Egorov sharply replied that the true "saboteurs" of Russian academic life were those who insisted on imposing one rigid ideology on everybody.

On December 21, 1929, Egorov was severely chastised at a meeting of graduate students at Moscow University. This was a new development: his own students were turning on him. They accused him of "religious zeal and proselytizing," of "ossification, inertia, lack of political zeal in reforming pedagogical research and methodology."[7] Egorov was greatly saddened to see his previous friends and students become his critics, but he absolutely refused to back down. He replied to the criticism by saying that his religious views were his own affair, and pointed out that instead of displaying "inertia" he had labored for years to improve mathematics in Russia—through the university, through the Moscow Mathematical Society, and in close collaboration with students who earlier, he noted, had expressed appreciation for his work on their behalf. In the spring of 1930 Egorov was dismissed by the university administration from his position as director of the Institute of Mathematics and Mechanics of the university, and replaced by the "Red professor" Otto Shmidt. (Kol'man had wanted the position, but was not considered adequately qualified; Shmidt, on the other hand, was both a Communist and a known mathematician of ability.)

In June 1930, still free and still a professor at Moscow University, Egorov attended the First All-Union Congress of Mathematicians in Kharkov. When he was asked at that meeting to sign a letter of greeting to the Communist Party Congress, taking place at the same time in Moscow, he pointedly refused, saying that the Moscow political meeting had nothing to do with his mathematics conference.

In September 1930 Egorov and more than forty other religious believers—some of them, like Egorov, members of the Name Worshipper circle—were arrested. They were accused not only of "mixing mathematics and religion" but also of "participating in a counter-revolutionary organization," the "Pure Russian Orthodox Church" and the "Catacomb Church." Egorov was kept in jail for a while in Moscow and then sent to prison exile near the city of Kazan on the Volga River.

Kazan is a city with a colorful and contradictory religious history. Over nine hundred years old at the time Egorov was sent there, it had originally been a major Tatar city, a Muslim area. In the sixteenth century, Tsar Ivan the Terrible of Moscow conquered the city and ordered all the Muslim mosques to be destroyed. Islam survived, however, and in the eighteenth and nineteenth centuries new mosques were built. By the early twentieth century, when Egorov ended up in Kazan, the city possessed a great variety of religious faiths and views—Russian Orthodox, Old Believer Orthodox, Muslim, Jewish, Lutheran, Catholic, and, in the new Soviet period, atheism. The two largest faiths were Russian Orthodox and Muslim, and the city was divided into "Russian" and "Tatar" quarters, with Lake Kaban in the center of the city dividing the two. This was a lake where, in the sixteenth and seventeenth centuries, Tatars were sometimes taken by Russians to be forcibly baptized into the Christian faith; if they refused, they were often drowned in the same waters. Even today, Tatars refuse to swim in this lake.

Recalcitrance about religion continued into the twentieth century. We are told that in exile Egorov continued to follow his religious beliefs, praying daily and practicing the Jesus Prayer. The prison guards persecuted him for doing so. In protest, Egorov refused to eat. Even before his arrest Egorov had had digestive problems, probably a stomach ulcer, and now those problems rapidly worsened. After weeks without food the 62-year-old mathematician was no longer able to stand, and his internal organs, particularly his liver, began to fail. The prison authorities sent him to a hospital in the Russian quarter of the city, on Butlerov Street, where a guard was stationed at the door to his room. The clinic was at that time called the "State Institute for the Improvement of the Qualifications of Physicians"

Fates of the Russian Trio

The hospital in Kazan where Maria Smirnitskaia cared for Egorov.

(GIDUV); today the same building houses a department of the Kazan State Medical Academy.

When Loren Graham visited this hospital in 2004, several people there said they had heard stories of Egorov's last days. Although one cannot be certain that they are true, they do reflect the way some people remember the event today. According to the stories, one of the physicians in the hospital was none other than Dr. Maria Smirnitskaia, wife of the mathematician Nikolai Chebotaryov, who had resigned his position at the Civil Engineering Institute in Moscow in 1924 when he learned that he had been given it because of the unfair dismissal of Egorov. Dr. Smirnitskaia remembered this episode and knew who Egorov was, and she did everything in her power to save his life. Unfortunately, the collapse of his internal organs had progressed to the point where nothing could be done. Nonetheless, Dr. Smirnitskaia was determined that the still-conscious Egorov should die in decent conditions. As the attending physician, she was qualified to sign Egorov's death certificate. She allegedly forged such a certificate while Egorov was still alive, gave a copy to the guard at the door to the hospital room, and suggested that he report to his superiors for reassignment. She assured the guard that the person he had been requested to watch was no longer living. After the departure of the guard, Dr. Smirnitskaia enlisted the willing assistance of her husband, Nikolai Chebotaryov, and together they loaded Egorov onto a hospital gurney, covered him with a sheet, and pushed him down the streets of Kazan to their second-floor apartment on Staro-Gorshechnaia Street (now Shchapov Street).

According to the stories, the next day Egorov actually did die, in the arms of Dr. Smirnitskaia, and his last words were those of Psalm 54, a phrase appropriate to his Name Worshipping creed: "Save me, O God, by Thy name!" ("Vo imia Tvoe spasi mia" in Church Slavonic).

Mathematicians in Kazan, at the local universities and several institutes, knew of the significance of Egorov's work and career. However, all but one—Nikolai Chebotaryov—were too frightened to attend Egorov's funeral, which was held in Arskoe Cemetery in Kazan, near the Yaroslavskie Chudotvortzi (Miracle Workers Church). The Arskoe Cemetery is the most prominent one in the city of Kazan, and it was at this time under the control of authorities who either refused, or feared, to allow a political prisoner to be buried in such an honored place. One way or another, however (probably through bribes given to the grave-diggers), Egorov was placed in an unmarked grave very near the tomb of one of Russia's greatest mathematicians—that of Nikolai Lobachevsky (1792–1856), founder of non-Euclidean geometry. Only many years later, after World War II, was it possible to place a tombstone on Egorov's grave, an action that was initiated by yet another mathematician, V. V. Morozov.

The bravery of the secular Chebotaryov in defending the religious Egorov had adverse effects on his own career. An outstanding mathematician, Chebotaryov was in 1938, 1943, and 1946 a candidate for election to full membership in the Soviet Academy of Sciences, a rare honor for a provincial scientist. However, the president of Kazan University and the head of the Communist Party organization of the university opposed his candidacy, citing his "reactionary" ideology.[8]

Fates of the Russian Trio

Egorov's gravestone, Arskoe Cemetery, Kazan.

The only evidence we have that might explain Chebotaryov's classification as "reactionary" by his superiors was his defense of Egorov. Thus the former Red Army soldier was unjustly treated because of his sense of fairness.

In 2004 Loren Graham visited Arskoe Cemetery in Kazan, located Egorov's grave, and photographed it. While standing in front of Egorov's tombstone, Graham was approached by a man playing a funeral dirge on a tuba. The tuba player, it seems, spent his days in the graveyard looking for mourners, and upon finding one he supplied appropriately sad music in the hopes of receiving a tip. As Graham gave him the expected reward, the tuba player looked at the tombstone and asked, "Did you know this man?"

"Not personally, but I know who he was, a famous mathematician," Graham replied. "Do you know who he was?"

"I have no idea who he was. Never heard of him. But I will tell you something: there is something very strange about this tombstone. The bottom bar of the cross slants in the wrong direction. According

{ 139 }

to Orthodox custom, the lower bar should slant from the upper left to the lower right when you are facing it. This one goes in the opposite direction. Whoever put that cross there knew nothing of religion."

In the meantime Pavel Florensky, the priest who had played such an important role in the discussions of mathematics and religion with Egorov and Luzin, was enduring his own ordeal with the Soviet police. In 1928 the police began an investigation of the Trinity Monastery of St. Sergey in Sergiev Posad, about 45 miles northeast of Moscow. This town was Florensky's home, where he lived with his wife, Anna Mikhailovna, and several children in a wooden house not far from the monastery. The monastery itself had served in the first years after the Communist revolution as a refuge for religious believers, including some Name Worshippers, and also for surviving members of the Russian aristocracy who hoped to find protection within its walls. The Soviet police regarded the monastery town as an unconquered bastion of supporters of the overthrown tsarist regime and its ideology.

The police investigation inspired a newspaper campaign criticizing the monastery. In May 1928, the *Workers' Newspaper* published several exposés. One journalist exclaimed, "All kinds of 'people of the past'—but mainly Grand Dukes, ladies-in-waiting, priests and monks—have built themselves a hive at the so-called Trinity-St. Sergey monastery." Another complained that the "revolutionary storm" had "hardly touched the centuries-old walls of this former citadel of depravity." This writer singled out the works of Florensky as "religious tracts," falsely described as "scholarship."

Four days later, on May 21, 1928, the police came to Florensky's house and arrested him. The warrant was signed by the head of the secret police of the entire country, Genrikh Yagoda. The police took Florensky to the Lubianka prison in central Moscow for questioning.

The investigation produced a few surprises. One of his questioners, a man named Poliansky, asked Florensky, "Have you ever been arrested before?" Florensky replied that he had, in fact, been arrested once before, by the tsarist police in 1906 for preaching a sermon in which he had protested the execution of one of the leaders of the unsuccessful 1905 revolution. Florensky did not sympathize with the revolution, but he opposed capital punishment. The discovery of Florensky's protest and subsequent arrest was an embarrassment for the Soviet police, who had falsely portrayed him as an unquestioning supporter of tsarism. The police then asked Florensky about his political beliefs. He replied, "I consider it harmful for society when scholars and scientists, whose vocation it is to be dispassionate experts, become involved in politics. I have never in my life belonged to any political party whatsoever."

And then his interrogators found out that Florensky was engaged in research for a Soviet military institution, whose administrators spoke well of his work. When the police asked him why, as a priest, he would work for the Soviet military, Florensky replied, "I took up the work voluntarily and suggested this field of investigation myself. I regard the Soviet authorities as the only real force capable of improving the conditions of the masses. I do not agree with certain measures taken by the Soviet authorities but am unconditionally opposed to any [outside, foreign] intervention, whether military or economic."9

Perhaps because of these facts, clearly news to the police, Florensky was given a relatively light sentence: three-year exile to the city of Nizhny Novgorod, where he was to report regularly to the police. Soon after his deportation to the city, Yekaterina Peshkova, former wife of the writer Maxim Gorky (after whom, in a curious twist of fate, the city of Nizhny Novgorod would later be renamed), appealed for Florensky's release, citing both his earlier arrest by the tsarist authorities and his loyal service to the Soviet government as a scientist. Another appeal was made by Ludwig Martens, chief editor of the *Technical Encyclopedia* and an old revolutionary. The appeals were successful, and after only a few months in Nizhny Novgorod, Florensky was allowed to return home to Sergiev Posad, where he resumed his work both with the Church and with the Soviet research institute. For a few more years, even after his former teacher Egorov and many others of his acquaintances had been arrested, Florensky lived in peace.

However, the priest-scientist continued his stubborn ways. Asked to deliver a technical paper to the Main Electrical Administration, he insisted on wearing his white priest's cassock while doing so, causing a sensation. Science and religion were being described in Soviet propaganda as inherent enemies, and here was a priest presenting scientific papers!

Florensky's next arrest came on February 26, 1933, and this time the police were much better prepared. By now they had perfected both their means of interrogation, using torture, and their accusations against those they arrested. The official description of Florensky by the police was now "a priest-professor, and an extreme right-wing monarchist in his political views." His interrogator, a man named Shupeiko, head of the political section of the Moscow secret police, charged him with being a member of a "counter-revolutionary party" attempting to overthrow the Soviet regime. Florensky had never heard of this organization, "The Party for the Rebirth of Russia," until he was charged with being one of its leaders.

Submitted both to torture and to dire threats to the lives of his family and friends, Florensky broke under the pressure. He signed a confession typical of those being forced upon political prisoners of the time: "Fully aware of my crimes against the Soviet system and the [Communist] Party, I wish to express in this document my profound repentance for my criminal membership in the nationalist-fascist center." Many years later, in 1958, during a time of rehabilitation of some of Stalin's victims, a Soviet court issued a decision stating that "Florensky (and other persons) were unjustly convicted without proof of their guilt."[10]

We now have access to some of the Soviet secret police archives about Florensky, but questions still remain about important details. For example, the records tell us that not only Florensky but also his mathematician friend Nikolai Luzin was accused of being a leader of the "nationalist-fascist center." Luzin, in fact, was accused of being "in charge of foreign ties" at the counter-revolutionary organization, no doubt because of his extensive connections with French and German mathematicians.[11] The secret police files even contain the ridiculous charge that Luzin had met with Hitler in Germany in order to receive espionage instructions.[12] However, we have no evidence that Luzin was arrested or threatened by the police at this time, so a mystery remains: why would such serious charges against Luzin not result in his punishment?

Once again, an effort was made to save Florensky from prison. Ludwig Martens wrote to Mironov, head of the secret police economic administration, saying:

> Professor Florensky is one of the most important Soviet scientists, and what happens to him will be of great significance for Soviet science as a whole, and for a great many of our research institutes. Being convinced that his arrest is the result of misunderstanding, I am appealing to you yet again to personally look into this case.
> With Communist greetings,
> Ludwig Martens[13]

This time, however, the appeal was unsuccessful.

In August 1933 Florensky was sent on a prison railway car to the Soviet Far East, to the town of Skovorodino on the Amur River, not far from China. At first deeply and understandably depressed, he soon recovered his interest in science and began studying local phenomena, such as permafrost. He also started to compile a dictionary of the language of a local Siberian ethnic group. However, his efforts were not appreciated by his prison supervisors. For reasons that remain unclear, he was soon transferred to one of the harshest prison camps in Soviet Russia, all the way back eastward to the Solovetsk Islands in the White Sea in the Arctic, arriving in October 1934.

The Solovetsk Prison Camp has an infamous place in the Gulag prison system of the Soviet Union. Established in the early 1920s, it was one of the first prison camps and became one of the most gruesome, a former monastery on a remote island where over half a million people are believed to have perished. In the midst of death and suffering, the many scientists, artists, and writers who were incarcerated there managed to establish study circles, music ensembles, and a theater. The Solovetsk camp was so notorious that after the fall of the Soviet Union, a memorial to those who died there, consisting of a red granite stone from the island, was placed near the headquarters of the KGB in Moscow, the Lubianka—a place where Egorov, Florensky, and many of their colleagues spent time en route to other camps.

In the labor camp Florensky once again found a subject for scientific study, this time the extraction of iodine and agar from seaweed. The effort was at first so successful that the prison camp operated a factory that processed seaweed, known as "the Iodine Enterprise." However, Florensky did not know that an informer was being confined with him in their prison cell, a man who attempted to engage him in political discussions in order to pass on to his superiors any incriminating remarks that Florensky might make. The informer, a man named Briantsev, reported that Florensky said the following in one of their conversations in the cell:

> In the Soviet Union they punish people for no reason at all. They kept demanding at the Lubianka that I name the people with whom I supposedly held counter-revolutionary conversations. After I had stubbornly refused to cooperate the interrogator said: "Of course we know that you don't belong to any organization and have not been carrying out any political agitation! But if something does happen our enemies could place their hopes on you. . . . We can't behave like the tsarist government and punish people for an already committed crime. Our job is to anticipate."[14]

For speaking in this way, Florensky was accused of "carrying out counter-revolutionary agitation" in the camp. The Iodine Enterprise was closed. Florensky was obviously headed for a tragic end similar to that of his old teacher, Egorov.

For many years the circumstances of Florensky's death were unknown, and even now questions remain. However, the recently available Lubianka archives contain two telling documents. One is a narrow strip of paper with "Florensky, Pavel Aleksandrovich" typed on one side, and on the other "To be shot," with a checkmark in red pencil. The other document states:[15]

> The death sentence passed on Florensky Pavel Aleksandrovich by the Leningrad Region NKVD Troika was executed on 8 December 1937.
> Commandant of the Leningrad Region NKVD
> Senior Lieutenant K. Polikarpov.

In October 2002 the Russian human rights group "Memorial" reported new evidence on the actual circumstances of Florensky's death. According to this information, in December 1937 Florensky was brought from the Solovetsk Islands to Leningrad, where for a while he was in a prison cell in the "Big House," the headquarters of the Leningrad secret police on Liteinyi Prospect (the building still exists and is still a police headquarters). Then, according to this new evidence, Florensky was forced to undress, his hands and feet were bound, and he was taken along with several hundred other people in a convoy of trucks to the Rzhevsky Artillery Range, near the town of Toksovo, about 20 miles south of Leningrad. There, we are told, they were all shot. Forensic scientists have found many thousands of skeletons there showing gunshots in the base of the skulls—a standard procedure of the Soviet secret police. Irina Fligye, head of the historical section of Memorial in St. Petersburg, reported on October 1, 2002: "There is a certain degree of indirect evidence that Florensky might have been executed in that area on December 8, 1937."

By 1930 Luzin had seen both his teacher Egorov and his friend and fellow student Florensky arrested by the secret police and imprisoned. He was already frightened about his own possible arrest, and now his fright turned into terror. His colleague, the mathematician A. Ia. Khinchin, described Luzin after 1930 in the following way:

"He feared for his whole life, he shook with fear. . . . This fear and trembling have remained with him until the present time [1936]."[16] Luzin wanted to be an honest man, a person who refused to compromise himself by making pro-Soviet statements in which he did not believe, but this became more and more difficult. The pressure under which Luzin was living is illustrated by a remark made to him by one of his older colleagues, an applied physicist by the name of Appel'rot, who had graduated from Moscow University in 1889: "Nikolai Nikolaevich, in our troubled times your task is to hold the candle of science up against the darkness of obscurantism, which you are doing. And because you are our leader in this effort, all the consequences will fall on you." After Appel'rot's statement Luzin just lowered his head and was silent.[17]

Under this kind of pressure Luzin's mental state deteriorated, and he spent long periods in sanitoriums trying to recover.[18] Somehow he managed to escape arrest, even though in 1933, in the investigation of Florensky and the so-called "nationalist-fascist center," Luzin had been accused of being one of the leaders of a "counterrevolutionary organization."[19] We can only speculate why Luzin was not arrested in 1933, if not earlier. One reason may have been that there was simply much less evidence against him, even in the eyes of the secret police. Egorov and Florensky had been much more open in showing their religious beliefs, while Luzin as early as 1922 began to conceal his inner convictions from the authorities. In 1929 Luzin stopped teaching at Moscow University and fled to the relative security of the Academy of Sciences, where he did not have to face undergraduates, many of whom were becoming increasingly radical. These students were often critical of the old professoriate inherited from the tsarist regime.

Just as Kol'man had helped bring down Egorov, now he set his sights on Luzin. In lectures and in various writings Kol'man castigated the Moscow School of Mathematics, which he saw as founded on idealistic and religious principles opposed by Marxist materialists. With Egorov gone, Luzin was the acknowledged head of the Moscow School of Mathematics and a logical target for Kol'man, who submitted secret denunciations to the police. One of these denunciations, dated February 22, 1931, has been found in the archives of the president of the Russian Federation.[20] In this document Kol'man criticized Luzin from the standpoint of intellectual Marxism in a way that may not have been entirely understandable to the police, most of whom knew little or nothing about philosophy or mathematics. The police understood very well what a "counter-revolutionary organization" was (since alleged membership in such an organization was one of their favorite and most deadly charges), but they were somewhat confused on the question of how Marxism should influence mathematics.

As a Marxist mathematician, Kol'man insisted that human knowledge finds its origin in the material world, not in the minds of scientists.[21] Marx and Engels had written that mathematics arose in the ancient world when humans found it necessary to quantify material things like olive oil and grain, and to measure land in primitive surveying operations. Thus, for Marxists, mathematics was a science of material relationships. According to Kol'man, while some areas of mathematics may have become very abstract in modern times, the discipline never lost its contact with the exterior world. He maintained that mathematics must be interpreted from the standpoint of philosophical materialism.

Opposed to this view, said Kol'man, was the "idealistic, religious" view that mathematics is merely created by human beings—that it is a product of their minds, without a necessary relationship to the material world. In a 1931 article Kol'man even used technical arguments about Luzin's treatment of the continuum, saying that Luzin "eliminates all points with rational coordinates, which has even less to do with reality than absolute continuity." Kol'man reproached Luzin for his "inability to understand the unity of continuous and discrete."[22] In his denunciation Kol'man accused Luzin of saying that numbers "exist as a function of the mind of the mathematician."[23] Here Kol'man was using the debate of twenty years earlier and Luzin's semi-intuitionist inclinations to accuse the latter of absolute idealism—the belief that a person gives a thing existence by thinking about it.

Kol'man, of course, explained these tendencies in Luzin's Moscow Mathematical School as reflecting the "pernicious influence of the bourgeois class" and imperialism. He carried the ideological fight outside the Soviet Union to the Second International Congress of the History of Science in London in 1931; there he gave, in the presence of Bukharin, a talk in which he adapted his accusations to the western European context by attacking Luzin's ideology and his allies in France, particularly Lebesgue. We now know that Kol'man had been appointed by the Soviet Communist Party to attend this congress in order to keep an ideological watch on the other Soviet participants, especially Hessen and Bukharin, who were under suspicion.[24] In criticizing Lebesgue and Luzin, Kol'man used arguments that had been developed by Borel against transfinite numbers. A French communist, Paul Labérenne, actively disseminated Kol'man's ideas in communist publications in France at the same time Luzin was facing threats to his life in Moscow.[25]

The philosophical issue that lay beneath this partisan discussion was of course an authentic one that has plagued thinkers since the time of Plato and Aristotle. On such questions Kol'man, however, left no room for nuance or subtlety. He was a true believer: mathematics was to him a matter of ideological faith. It is one of the tragedies of Soviet history that sometimes legitimate philosophical issues like the age-old opposition of philosophical idealism to realism or materialism were converted into lethal weapons in a battle that sometimes resulted in the deaths of the defenders of idealism. Kol'man was a militant Marxist leader in those battles.

Kol'man expressed these views in lectures at the "Red Professors' Institute of Philosophy" and at Communist Party ideological meetings. He was supported by other militant Marxist philosophers, including V. Molodshii of the Institute of Philosophy of the Academy of Sciences[26]—the person who as a young student had been offended

Fates of the Russian Trio

by Egorov's refusal to discuss mathematics with him because he was a member of the Young Communist League.

These accusations of philosophical and ideological sins, serious as they were, did not in themselves provide sufficient reasons to topple the world-famous Nikolai Luzin, one of the Soviet Union's most prominent mathematicians. Personal and practical factors would have to come into play before the attacks of Kol'man and his friends could be successful. And those other factors now entered in.

By the early 1930s Lusitania, the informal organization of mathematics students centered around Luzin at Moscow University, no longer existed. After all, Luzin was no longer teaching at the university, confining himself to research duties. Even more important, Luzin's former students were no longer willing to treat him as their revered master. A number of them had become world-famous mathematicians in their own right, and several of them were creating "schools" of their own, including Alexandrov, who established modern topology in opposition to the old descriptive set theory. Furthermore, several of them were jealous of Luzin. They accused him of borrowing some of their ideas and discoveries and using them as his own. (When a professor and his or her students work out ideas together it is frequently difficult, if not impossible, to assign credit correctly.) Several of these younger mathematicians also resented the fact that Luzin still chaired important committees that controlled the granting of degrees and the promoting of mathematicians in various institutions. For example, as a member of the prestigious Academy of Sciences, Luzin could block the election of younger mathematicians hoping to ascend to this pinnacle of Soviet science. Luzin's removal from these positions of influence might open up avenues of advancement for younger mathematicians, including some of his former students. A struggle for power between the generations was shaping up. Finally, a number of the younger mathematicians were much more supportive of the Soviet order and even of Marxism than Luzin was. The situation that was developing was not a pretty one. Some famous Soviet mathematicians, whose names are still well known in the world of mathematics, would join in a ritualistic denunciation of their former teacher, trying to get him out of the way of their professional advancement. The campaign against Luzin contained disparate elements, with people pursuing various different goals. Kol'man, as we have seen, was a true ideologue, dedicated to the defense of Marxist interpretations of mathematics and therefore inherently opposed to the religious Luzin. Some of Luzin's former students who were now professors themselves, especially the topologist P. S. Alexandrov, were not interested in Marxist philosophical analyses but were simply jealous of Luzin and wanted to get him out of the way. For militant young undergraduates, members of the Komsomol, it was enough to know that Luzin was a member of the old "bourgeois" pre-revolutionary professoriate in order to oppose him. Finally, in the increasingly patriotic Soviet atmosphere after the mid-thirties (Hitler had come to power in Germany in 1933), Luzin's close ties with mathematicians in western Europe and the fact that he often published papers in foreign journals signaled to nationalist critics that his devotion to his native land was questionable.

Luzin was aware of these threats against him, and he tried to defend himself. In an effort to show his loyalty to the Soviet order, he wrote articles in the field of applied mathematics that might be useful for Soviet industrial and military efforts. He visited elementary and secondary schools and helped foster an interest in mathematics among young students; he found these children safer as an audience than undergraduates or graduate students politically attuned to Soviet propaganda. Although he no longer taught at the university, he nonetheless participated in dissertation examinations and admission committees. He bent over backwards to praise the new Soviet system of education and the young students from working-class and peasant families who were entering the universities.

Many of these students were poorly prepared and could not master the mathematics which their professors were teaching. But Luzin knew that if he criticized their performance he would be vulnerable to the charge that he had a class bias, or even that he was "anti-

Soviet." He would not bend his principles in the universities by promoting mathematicians of inferior quality, but he thought that at the grade-school level such standards were not so important, and therefore he sought places where he could praise Soviet students without doing much harm to his profession, even when he knew their work was inferior. It was this unjustified praise that gave his enemies the opportunity they were seeking in order to attack him.

The anti-Luzin campaign that soon developed shows signs of being organized not by the police or the Communist Party, but by Luzin's personal enemies on a lower level. Several post-Soviet writers in Russia who have looked back on these events have used the words "intrigue" or "conspiracy" to describe the movement against Luzin. A trap was set for Luzin, and the main organizer was his inveterate enemy, Kol'man.

Kol'man knew that Luzin sometimes visited local secondary schools in the Moscow area in order to encourage mathematics education, and he suggested to a reporter at the newspaper *Izvestiia* that there might be "a story" about Luzin's visit to a trigonometry class in School No. 16 in the Dzerzhinsky Region of Moscow. The reporter went to the event and afterwards asked Luzin if he would "share his impressions" with the readers of *Izvestiia*. Not suspecting a thing, Luzin agreed, and wrote a short article praising the class in exuberant terms which was published in the paper on June 27, 1936. He wrote that he was amazed by the quality of the class, saying that when he asked more and more difficult questions, the correct answers were always given. He added that he "was not able to find any weak students in the class."

This was the opening Kol'man was hoping for. By giving false praise of a trigonometry class that actually contained some poor students, Luzin was now vulnerable to the charge of "wrecking," a deadly serious charge of sabotage against the Soviet order. Kol'man believed that he could equate the false praise of mathematics students in the schools to the purposeful wrecking of industrial production by throwing a monkey wrench into a turbine—a charge made earlier about anti-Soviet engineers. Kol'man followed up Luzin's visit to the school with a vicious article in the Party newspaper *Pravda* entitled "On Enemies Hiding Behind a Soviet Mask," in which he accused Luzin of trying to harm Soviet education by intentional praise of inferior work. Luzin knew very well, Kol'man wrote, how false his published evaluation of the class was and actually "joked" about it with his closest friends. (This may actually have been true, but we have no way of knowing for sure.) And then Kol'man made the sort of accusation against Luzin that usually ended with arrest and imprisonment:[27]

> We know in what way Luzin grew up. We know that he is a member of the inglorious tsarist "Moscow Mathematics School" whose philosophy is one of right-wing reaction based on religious orthodoxy and autocracy. We know that even now his views are not far from these origins, perhaps a little "modernized" in a fascist way. . . . Luzin has remained an enemy concealed behind an impenetrable mask of social mimicry which he has pulled over his face.
>
> You won't get away with it, Mister Luzin! Soviet science will rip away your mask!

Kol'man had made an alliance with L. Z. Mekhlis, the editor of the Party newspaper *Pravda*, in order to "unmask" and overthrow Luzin. On July 3 Mekhlis wrote a letter to the Central Committee of the Communist Party, headed by Stalin himself, asking for an investigation of the "situation in Soviet scientific institutions" signaled by the "Luzin Affair." Stalin, who knew of Kol'man and considered him to be a self-promoting intriguer, was not immediately enthused. Nonetheless, he sent a note to his assistant Molotov (whose duties included supervision of the Academy of Sciences) with the somewhat casual remark, "It seems we can go ahead with this investigation."[28] Mekhlis then instigated, through the pages of *Pravda*, a campaign of public denunciation of Luzin.

Meetings were held in various scientific institutions—the Steklov Mathematics Institute, Moscow University, the Institute of Energy, Leningrad University, the Belorussian Academy of Sciences, and

Fates of the Russian Trio

others—which ended up issuing "proclamations" denouncing Luzin's perfidy. In response to this outcry, the presidium of the Soviet Academy of Sciences set up a special investigative commission headed by one of its vice-presidents, Gleb M. Krzhizhanovsky, and containing many of Luzin's academic colleagues (eleven in all). Among the eleven were three younger mathematicians, all former students of Luzin, who were known to be his rivals and who were not well disposed toward him—P. S. Alexandrov, L. G. Shnirel'man, and A. Ia. Khinchin. Two others, O. Iu. Shmidt and S. L. Sobolev, were active members, respectively, of the Communist Party and the Komsomol, and could be expected to agree with whatever line the Party ended up supporting. The positions of the most influential administrators in the Academy of Sciences—Krzhizhanovsky himself, N. P. Gorbunov, and A. E. Fersman—were not so clear. At first they demonstrated uneasiness about leveling charges against Luzin that would almost certainly result in his imprisonment and perhaps in his death. They seemed to favor some sort of reprimand that would permit Luzin to continue his scientific research. Only one member of the commission, the older mathematician S. N. Bernshtein, openly tried to defend him. Two others, I. M. Vinogradov and A. N. Bakh, spoke rarely and probably secretly sympathized with the beleaguered mathematician.

The commission conducted a ten-day full-scale interrogation of Luzin at sessions in which most of the leading mathematicians of Moscow participated—either as members, as witnesses, or as part of the attending audience (members of which were also encouraged to criticize Luzin). Great pressure was put on all of Luzin's associates to participate in one way or another, to lash out at their colleague and teacher. A conspicuous absentee was Nina Bari, one of his former graduate students, now an outstanding mathematician herself (in 1926 she received a state prize for her work on trigonometric functions). She was rumored to be Luzin's lover, and she flatly refused to come to the meetings where the man she revered and adored was being flailed. In Luzin's presence, one of his interrogators referred to her sneeringly as "a person devoted to you, and I will not

Nina Bari.

say more than that." (Many years later, after Luzin's death, and after she had finished publishing his collected works, Nina Bari would commit suicide, like Anna Karenina, by throwing herself in front of a train—in Bari's case, the Moscow subway.)

In the first day or two of the commission's work a few of Luzin's colleagues, particularly Bernshtein, tried to defend him, but it soon became clear that no real defense was permitted. As the days passed the attacks became more and more vicious, with his colleagues calling him an "enemy of Soviet power" and a "wrecker," terms which in the context of the time amounted to capital crimes.[29] It was demeaning and tragic that outstanding mathematicians like Alexandrov, Khinchin, Sobolev, Kolmogorov, Liusternik, and Pontriagin—who knew in their hearts that Luzin was not an active opponent of the Soviet Union, much as he may have disagreed with some Soviet policies—all agreed that their colleague was a traitor. Several of them even invoked the secret police, implying that the case was so serious that these punitive authorities might have to take action against Luzin.[30] And they combined these deadly political comments with accusations that, as a mathematician, he stole his students' ideas (that is, their own ideas). They further accused him of sending his best articles abroad for publication (to France or Germany) while publishing only "inferior ones on applied mathematics" in the Soviet Union. (There may have been some truth in this charge, since the most prestigious mathematics journals in the world at this time were in western Europe and more inclined toward pure mathematics, and it was a common practice for ambitious mathematicians to try to publish there.) In the increasingly nationalistic atmosphere of the Soviet Union, this publishing pattern became a heavy criticism. The most aggressive attacks against Luzin were delivered by Alexandrov, his former student, who was fiercely jealous of his teacher.

Many times during the investigation the names of Luzin's French colleagues—Borel, Lebesgue, Baire, Denjoy, and others—were mentioned, both because they worked on the same topics in mathematics as Luzin and also, more ominously, because Luzin's connections with them were used by some of his critics to impugn his loyalty to the Soviet Union (such as the exaggerated friendship with Borel and long-lasting ones with Denjoy and Lebesgue). How strange it must have seemed to Luzin to face his critics in Moscow while thinking about his past visits to France, the deep conversations with his friends there about the mathematical problem of the continuum, often held over good food and wine. Paris seemed at that moment a universe away, but one to which he was still drawn. He deeply admired French culture, and had once asked in a letter to Otto Shmidt, "What does Paris give one? It gives literally everything."[31] Especially memorable for Luzin were the touching and romantic times in France with his adoring student Nina Bari, who also received a Rockefeller grant, and whose time in France overlapped with Luzin's. There were also visits by Luzin with his wife to a lovely country house on the island of Oléron in Brittany, where the Denjoys hosted them in the summer.

As Luzin faced his inquisitors he knew Nina was in Moscow, at the university, and he knew why she did not come to the meetings of the commission. Both she and Luzin realized of course that his life was in danger, in part because of his connections with France. Kol'man in his denunciation had mentioned the fact that Luzin had been a guest at the home of the mathematician Émile Borel at the same time that Borel was Minister of the Navy in the French government. (The denunciation was made in 1931, but in 1936, at the time of Luzin's ordeal in front of his colleagues, Borel still worked with the French Navy.) To some of Luzin's critics, his ties to a "foreign militarist" like Borel seemed evidence of disloyalty.

Luzin probably feared that word of his difficulties in Moscow would get out to his French friends and cause them to try to help

The Luzins with the Denjoy family on the island of Oléron, Brittany, c. 1930. (left to right): Nadezhda Luzin, Nikolai Luzin, Fabrice Denjoy, Arnaud Denjoy, and a family friend, Madame de Ferrières, with her son. The two boys are still alive.

Fates of the Russian Trio

him, but that doing so would only deepen the suspicions of his critics. His French colleagues did, indeed, soon learn of Luzin's difficulties, but they were aware that a loud and public outcry in France, arising from published petitions of mathematicians, might actually harm Luzin's case. Therefore, they chose to act privately. They submitted a confidential letter supporting Luzin to the Soviet ambassador to France, V. P. Potemkin. That letter was delivered to the Soviet embassy in Paris on August 13 by Borel and Paul Langevin.[32] Several left-wing mathematicians, most notably Jacques Hadamard, closely connected to the French Communist Party through his daughter, did not sign the letter, evidently refusing to speak against the Soviet Union. (Langevin, who did sign the letter, would later join the French Communist Party himself, but several years after this event.) André Weil, who knew Alexandrov from Göttingen where he had met him in the summer of 1927, refused to sign the letter on what he called a strictly mathematical basis. Weil (as in other circumstances) was naive about politics and here failed to understand the human situation behind the conflict between two mathematical schools. In the end, the French letter to the Soviet ambassador had no effect because by the time it was received, the Luzin affair in Moscow had reached its conclusion.

By the third or fourth day of the investigation, it seemed clear to everyone that Luzin was destined for arrest, imprisonment, and possible execution. But behind the scenes something surprising was happening, something that would not be fully revealed for more than fifty years. On July 6 the well-known physicist Peter Kapitsa (later a Nobel laureate) wrote a confidential letter to Molotov, who delivered it to Stalin, in which he made the case for not harming Luzin. Kapitsa said that he had no idea whether the charges against Luzin were true or not, but that Luzin was so valuable to the Soviet Union as a mathematician that his talents must be used for the good of the country. He went on to observe, "[Isaac] Newton, who gave us the law of gravity, was a religious maniac.... [Girolamo] Cardano, who gave us great mechanical and mathematical discoveries, was a drunk and a debaucher.... What would you do with them if they lived in the

{ 157 }

Soviet Union?"[33] Answering his own question, Kapitsa urged the leaders of the Soviet Union to preserve such people in specially guarded circumstances where they could do no harm, but where they could contribute to Soviet strength.

Kapitsa knew what he was speaking of. He had essentially been kidnapped by Stalin in 1934 and forced to work in the Soviet Union instead of Britain, where he was living at the time and where he had planned to live permanently. During the years in Moscow that followed, and for thirty years after that, Kapitsa regularly sent confidential letters to the leaders of the Soviet Union with advice which was—amazingly—frequently followed. In all, he wrote 45 letters to Stalin, 71 to Molotov, 63 to Malenkov, and 26 to Khrushchev. When the secret police arrested his colleagues Vladimir Fock and Lev Landau, both distinguished physicists, Kapitsa sent letters to Stalin defending them. Now he did the same for Luzin. All three men would escape the long detentions in prison for which they seemed destined.

There is no known case of another Soviet intellectual who was able to speak so frankly to the leaders of the Soviet Union and get away with it. There was something about Kapitsa that intrigued these leaders. They probably recognized that Kapitsa was a truth-teller who was, ultimately, harmless to the Soviet regime because he kept his appeals confidential. He did not organize resistance or ask others to join him in his campaigns. Kapitsa did not reveal his correspondence with Soviet leaders to others (even his secretary did not know about the letters), and that correspondence did not become publicly known until after Kapitsa's death in 1986. At one point Lavrentii Beria, head of the secret police from 1938 to 1953, threatened to arrest Kapitsa, and Stalin sent him a note saying, "I'll take care of him personally. Don't touch him."

At the same time that Kapitsa was making his appeal to the Soviet leaders, the head of the investigatory commission, Gleb Krzhizhanovsky, also had his own doubts about the dire direction in which the commission was going. He seems to have been abashed, and perhaps even a little horrified, at the way in which Luzin's colleagues maneu-

Peter Kapitsa.

vered to get rid of him so that their own careers would be advanced. Krzhizhanovsky was an old revolutionary with personal connections to Stalin, and he decided to talk to the top leader himself about the Luzin affair.

Krzhizhanovsky and Stalin had a conversation about the matter, probably on July 11 or 12, 1936. No record of the conversation has been found, but it is clear that the two men decided that Luzin's colleagues were engaged in an excessive campaign against him. Stalin had not, in the beginning, been enthusiastic about the investigation of Luzin because he had his own reservations about Kol'man. Perhaps the correct path, Stalin and Krzhizhanovsky thought, was to do as Kapitsa had recommended—to reprimand Luzin severely, keep him in conditions where he could do no harm, but allow him to continue his work.

On July 13 Krzhizhanovsky returned to the commission and announced that a few changes must be made. "We have been advised," said Krzhizhanovsky, that the charge that Luzin was an enemy of the state and intentionally tried to harm the Soviet Union should be removed. (This was, of course, the most deadly accusation; everybody knew what "we have been advised" meant.) Instead, Krzhizhanovsky said, Luzin should be described as a scientist who has engaged in activity "not worthy of a Soviet scientist"; he should be warned to mend his ways; and he should be removed from positions of administrative influence. Thus, Stalin and Krzhizhanovsky made a concession to Luzin's colleagues who wanted him out of the way of their own careers, but at the same time they followed Kapitsa's advice that even "religious maniacs" should be given a chance to contribute to the strength of the Soviet Union in science.

It was a surprising development, not at all consistent with some of the other things Stalin was doing and would do in the future. After all, in the bloody purges that were just getting under way, many outstanding scientists and generals important to the Soviet Union's war preparedness would perish (for example, the outstanding military theorist Marshall Tukhachevsky was executed less than a year later). But Stalin, with advice from Kapitsa and Krzhizhanovsky, seemed to believe that this massive attack on Luzin "from below" should not be allowed to come to its logical conclusion. Maybe Stalin wanted to show that he was the one in control of purges, not those who made denunciations, or maybe he really believed that Luzin could contribute to the war preparedness of the Soviet Union as it faced Nazi Germany (there is little evidence, however, that Luzin's work had any significant results for defense). Some scholars have theorized that Stalin did not want to raise the alleged (and ridiculous) connection of Luzin with fascism and Hitler because Stalin was already contemplating the possibility of an alliance with Hitler, as happened three years later in the "Hitler-Stalin Pact." But these are speculations. We will never know the inside of Stalin's mind. All we do know is that Luzin was saved from imprisonment and death. Luzin promised to rectify his flaws, and after 1936 he almost completely abandoned foreign publications.

Luzin never forgave his former student Pavel Alexandrov for what he had done to him in 1936. Although Alexandrov was a very distinguished mathematician, he did not become a member of the Soviet Academy of Sciences until 1953, more than two years after Luzin's death. Circumstantial evidence indicates that as long as Luzin was alive, he quietly blocked his nemesis from membership in the Academy—perhaps the highest honor in Soviet science.

8

Lusitania and After

> "I have been waiting for you for a long time; I just did not know what your name would be."
>
> —*Nikolai Luzin's response to 15-year-old Lev Shnirel'man when the unknown youth showed him some of his mathematical work*

FOR MANY YEARS a visitor to the Mathematics Department of Moscow University could see posted on one of the bulletin boards a genealogical chart of the "Moscow School of Mathematics," showing more than a hundred of its members. At the top of this chart was Nikolai Luzin, often considered the father of the Moscow School of Mathematics. During the Soviet years, Egorov's name did not appear on the chart because of his arrest, detention, and subsequent death. Even after the Soviet Union collapsed, Egorov was often not given the credit he deserved, perhaps out of habit.

The chart showed not only the names of many of the mathematicians in this famous school but also several of its branches, a result of the fact that a number of Egorov's and Luzin's former students soon had students of their own who became influential in the mathematical world. In our reproduction of this chart below, we have added the name of Egorov.[1] After all, he was Luzin's teacher, and he deserves a place at the very top of the chart. Some of the people listed on the

Lusitania and After

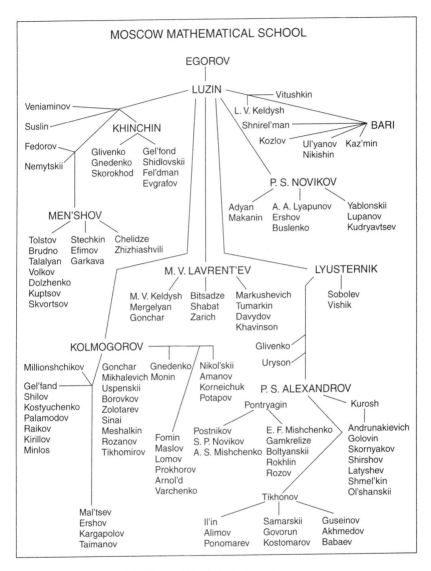

Genealogical chart of the Moscow School of Mathematics.

departmental chart as the students of Luzin could just as accurately be described as students of Egorov. Since both Luzin and Egorov were active in the famous Lusitania group, quite a few students owed their lineage to both professors.

Among the individuals on this chart are some of the most famous scientists in the Soviet Union, including a leading theoretician of the Soviet space program and president of the Soviet Academy of Sciences (M. V. Keldysh); the founder of the Siberian branch of the Academy of Sciences (M. V. Lavrent'ev); a brilliant analyst and applied mathematician who was a leader of the atomic weapons program (S. L. Sobolev, who actually did not move to Moscow from Leningrad until 1934); and many other prominent scientists and administrators. Among those whose names are best known in the history of mathematics itself are Andrei Kolmogorov (often described as one of the three or four most distinguished mathematicians of the twentieth century), P. S. Alexandrov, P. S. Novikov, L. S. Pontriagin, I. M. Gel'fand, and V. I. Arnol'd. One of the men on this chart, Ya. K. Sinai (a former student of Kolmogorov, now at Princeton University), has recently edited two books that tell important parts of the story of the Moscow School of Mathematics.[2]

What this chart does not show, of course, is the human side of these people—their personalities, relationships, or personal fates. But they were human beings like everyone else, with passions, strengths, and weaknesses. There are women as well as men on the chart; the former include Nina Bari and Ludmila Keldysh. Keldysh, together with Luzin, Bari, and other members of Lusitania, worked together with Sierpinski in Warsaw (to which he returned in 1919) and contributed a fascinating new chapter in the history of descriptive set theory. Her activities in the field lasted until World War II, partly in collaboration with her husband, P. S. Novikov, also listed on the genealogical tree of Moscow mathematicians.

Several of the mathematicians on the chart, like Novikov and Keldysh, married each other; some of them had love affairs and adulterous relationships with each other, both heterosexual and homosexual; some were religious believers, and some were atheists; some

Ludmila Keldysh.

were Communists, and some were thoroughly opposed to Communism; some of them were tolerant and wonderful people, but some were anti-Semites and dogmatists. We have seen how careerism sometimes destroyed friendships among former Lusitanians. Some of them died natural deaths, while some had violent deaths; several of them committed suicide; some died very young; some lived to old ages; and some are still alive today. Quite a few were Christians; some were Jewish; several had Muslim backgrounds; and a few could find Buddhists among their ancestors. In sum, it was a heterogeneous group—religiously, ethnically, and personally.

The chart contains a number of individuals who were the "fathers" or "mothers" of schools of mathematics of their own, whose students went on to make significant mathematical achievements. Thus the original school branched out in many directions: topology, groups and algebra, functional analysis, and so on. Leaders in this branching include (besides Egorov and Luzin) Khinchin, Bari, Novikov, Men'shov, Lavrent'ev, Lyusternik, Kolmogorov, Alexandrov, Pontryagin, Kurosh, and Tikhonov. Other branches could be added, particularly for the achievements in recent years, although the chart would then include students in other countries, such as Israel, the United States, France, and Germany.

Obviously we cannot discuss all the mathematicians on this chart; even describing the lives and achievements of the leading figures would require a separate book and would almost amount to a history of recent Russian mathematics. In this chapter we will tell the stories of several of the early members of Lusitania whose lives were especially closely intertwined with the political and social environments of their time.

Several outstanding mathematicians included on the chart died very young; if they had lived, it is likely that they would have become leaders in the field. One thinks of Uryson, who died at the age of 26 in a tragic accident while swimming with Alexandrov; of Suslin, who died from typhus at the age of 25; and of Shnirel'man, who committed suicide at the age of 32. All three were brilliant mathematicians.

We will look first at the talented and tragic life of Lev Genrikhovich Shnirel'man (aka Schnirelmann), who was born in 1905 in Gomel, where his father was a teacher of the Russian language. He excelled in school and already by the age of 12 was demonstrating an extraordinary mathematical ability, studying advanced texts. Although there may be apocryphal elements in this story, we are told that in 1920, at the age of 15, he went to Moscow and attempted to enroll in Moscow University in order to study mathematics. The minimum age for entering the university was 16. Shnirel'man somehow managed to arrange an appointment with Luzin, to whom he displayed a notebook in which he had written down his attempts to solve difficult problems. As the story goes, Luzin examined the notebook, looked up at the adolescent standing before him, and said, "I have been waiting for you for a long time; I just did not know what your name would be." Luzin also told Shnirel'man that he had recently had a dream in which a youth came to him who could solve the Continuum Hypothesis. Luzin went to the university authorities and obtained permission for Shnirel'man to enroll, and he became a member of Lusitania.

Shnirel'man finished his undergraduate work in two and a half years and immediately entered graduate school in mathematics, working under Luzin's direction. After completing his graduate work he spent a year in Novcherkassk as a young professor of mathematics and then returned, in 1930, to Moscow, where he spent the rest of his life.

In just a few years Shnirel'man made significant contributions to topology and number theory. Working with another Lusitanian, Lazar Lyusternik, he developed what is today called the "Lyusternik-Schnirelmann Category," an important new invariant defined for all topological spaces. He also made the first breakthrough on Goldbach's conjecture, which claimed that any even number is the sum of two primes;[3] Shnirel'man established the first result in this direction by proving that any natural number greater than 1 can be written as the sum of not more than 300,000 prime numbers. In recognition of his achievements, Shnirel'man in 1933 was elected a corresponding member of the Academy of Sciences, at the remarkably young age of 28.

On September 24, 1938, at the age of 32, Shnirel'man committed suicide in his Moscow apartment by closing all the doors to the kitchen and turning on the gas in the stove. The circumstances are still not entirely understood, but we have been told that shortly before his death Shnirel'man had been arrested by the secret police and submitted to an interrogation. During the investigation he was shown a fabricated confession (actually a denunciation of several of his friends) and was coerced into signing it. Torture was probably a part of the procedure, although we do not have this evidence. At any rate, Shnirel'man signed the confession and was released. Upon returning to his apartment he descended, understandably, into a deep depression. He told his close friend and mathematical associate, Lazar Lyusternik, that he had done "something very bad under pres-

Lev Shnirel'man.

sure." Lyusternik could guess what had happened. At that time the secret police frequently demanded such signed denunciations and then later showed these documents to other detainees in subsequent interrogations, "proving" that a person's friends and colleagues had accused them of awful crimes.

What did Shnirel'man say in his denunciations? Did he criticize Luzin, his teacher? Did he denounce his colleagues, some of the most brilliant mathematicians in Soviet history? Perhaps someday the still-secret archives will reveal the answers to these questions.

When Lusitania was formed, before the Russian Revolution, the moral atmosphere in Russia was rather permissive, and this continued in some circles for quite a few years after the Revolution. Even though homosexuality was banned in tsarist Russia (as it would be much more strictly in the Soviet Union), tolerance for homosexuality was common in large cities during the last years of the tsarist empire. European, and especially German, homosexual movements found a response in Russia during the last decades of tsarism. The composer Tchaikovsky's homosexuality was known in Russia at this time, and a number of Russian musicians, poets, writers, and mathematicians were similarly inclined, sometimes openly. In general, the period before and immediately after the Russian Revolution of 1917 was a time of sexual permissiveness. A prominent friend of Pavel Florensky's, Vasilii Rozanov (1856–1919), propagated a vision of "divine sexuality." Two of the leading Symbolist writers, Dmitrii Merezhkovsky (1865–1941) and Zinaida Gippius (1869–1945), lived in a *ménage à trois* which they considered a form of "embryo church."

Florensky himself, with his close connections to Luzin and Egorov, included among his unconventional views a praise of dyadic friendships, which were usually same-sex unions. Richard Gustafson wrote in his introduction to the English translation of Florensky's *The Pillar and Ground of Truth* (first published in Moscow in 1914, about the same time Lusitania was formed) that the book was to his knowledge "the first Christian theology to place the same-sex relationship at the center of his vision." Such views, in addition to Florensky's Name Worshipping practices, were one more reason why his teachings could not be accepted by the establishment of the Russian Orthodox Church. And the issue is still alive today. A recent look at the "Russian National Gay, Lesbian, Bisexual and Transexual Website" shows that Florensky is regarded as a founding figure.[4]

From its earliest days Lusitania was associated with several religious, ideological, and sexual views that would later, as the Soviet Union became more and more repressive and morally conservative, become literal time bombs. We have seen that in its origins there were significant religious elements. It was founded by representatives of the old pre-Revolutionary intelligentsia who were unsympathetic—even hostile—to the coming Soviet order. And it included among its early members homosexuals who would later encounter the weight of official repression and castigation. In view of these coming events it is not surprising that eventually Lusitania would fracture, its members turning against each other. Nor is it surprising that militant Soviet revolutionaries would eventually attack Lusitania and all it stood for.

Two of the early male members of Lusitania, P. S.Alexandrov and A. N. Kolmogorov, later formed a homosexual union, and a third, P. S. Uryson, was almost certainly homosexual as well. All three of these men were very talented mathematicians whose names will long be preserved in the history of mathematics. Alexandrov and Kolmogorov became world-rank mathematicians, each of them founding his own school of mathematical thought; and the same would doubtless have been true of Uryson if he had not died at the age of 26. Kolmogorov was one of the three most significant mathematicians of the twentieth century, along with David Hilbert and Henri Poincaré. Alexandrov, together with Fréchet, Hausdorff, and others, was the creator of modern topology.

In his younger years Alexandrov seemed not to be certain of his sexual orientation, or perhaps he was bisexual. The reminiscences about his personal relationships in his writings and correspondence, while not explicit, are surprisingly frank, as shown in his "Pages from an Autobiography."[5]

Pavel Sergeevich Alexandrov was born in Bogorodsk but grew up in Smolensk, a city about 250 miles west of Moscow. His father was a distinguished surgeon from a wealthy merchant family, and both his parents were well educated; Pavel grew up in privileged surroundings, learning French and German as a child. Very early he displayed interests that would remain with him throughout his life: music, the theater, swimming, and mathematics. His brothers were talented musicians, and his home was filled with music; Alexandrov often read or did his homework while his brothers played the piano or the violin. In later years he regularly attended concerts and theater performances, sometimes almost every evening. Swimming became another of his passions. In his memoirs he emphasized what an important event his first dip in the Dniepr River was, at the age of five. As an adult he would swim almost every day in the spring, summer, and fall, often with one of his mathematician friends, and often

Lusitania and After

in the nude. The swims almost always took place in rivers and lakes, not in institutional pools.

In Smolensk Alexandrov attended a classical gymnasium where he encountered mathematics, a subject which entranced him. His first teacher was Aleksandr Eiges, a person who remained important to him until Eiges's death in 1944. At one moment, years after that first meeting, Alexandrov thought he was in love with Eiges's sister, and went so far as to marry her, but that attempt turned out to be, by his own account, "a disaster."

In Alexandrov's description of Eiges's lectures on mathematics, one can see characteristics that would later show up in his own mathematical creativity:

> Geometry interested me even more than equations did, because in geometry there were axioms and theorems and proofs, and not only problems.
>
> When we came to the theory of parallel lines, Eiges began with amazing pedagogical tact and skill to tell us about Lobachevsky's [non-Euclidean] geometry. The very statement of the problem astounded me. Never before had anything aroused my interest and enthusiasm to that extent. Geometry became an enchanted kingdom for me, and I dreamed of that alone.

In later years Alexandrov would, together with his intimate friends Pavel Uryson and later Andrei Kolmogorov, develop important parts of modern topology.

In 1913 Alexandrov graduated from the gymnasium in Smolensk and immediately entered Moscow University. Thus he was present at the birth of Lusitania. His description of Luzin, his professor, is striking:

> To see Luzin in those years was to see a display of what is called an inspired relationship to science. I learnt not only mathematics from him, I received also a lesson in what makes a true scholar and what a university professor can and should be. Then, too, I saw that the pursuit of science and the training of

young people are two facets of one and the same activity—that of a scholar.

The relationship of the two men did not end well, however; as we saw in the previous chapter, Alexandrov later turned viciously on Luzin, even endangering his teacher's life by joining colleagues in calling him a traitor to the Soviet Union in a public tribunal being watched by the secret police.

Alexandrov achieved his first important mathematical result, under Luzin's tutelage, in 1915, when he was 18 years old. He proved that every non-denumerable Borel set contains a perfect subset, thus proving the Continuum Hypothesis for Borel sets.[6] Luzin was deeply impressed by Alexandrov's feat, and realized that this adolescent in his class possessed a truly great mathematical talent. Luzin then assigned Alexandrov an unsolved problem in set theory, which the latter was unable to solve. Alexandrov was so disappointed by this failure (he called his efforts "a serious catastrophe") that for a while he thought he was unfit for mathematics and should abandon the field. He left Moscow and entered the musical and theatrical worlds of Chernihiv (Chernigov), a city in northern Ukraine. He became a theater producer, and consorted mainly with musicians, poets, and artists. One of Alexandrov's major theater productions was Ibsen's *Ghosts*.

At this time a civil war was going on in Russia between the White and Red armies. Chernigov was briefly occupied by the White army of General Denikin, and Alexandrov was then arrested and accused of "actively and energetically collaborating with the Bolsheviks and thus supporting Soviet power and contributing to its popularity." The details of this arrest are sketchy. We have no evidence, not even from Alexandrov himself, that he had been politically active in any way; he tells us only that he had given public lectures on the arts. He wrote about his arrest during the later Soviet period, when a record of trouble with the Whites was a sign of merit; at this time he obviously took pride in this brief "criminal" past.

Lusitania and After

It seems likely that there is much more to this story. Ibsen was a critic of capitalist excess in a way that pleased neither its apologists nor its radical critics (Friedrich Engels, Rosa Luxemburg, and Anatoly Lunacharsky all criticized Ibsen at one time or another). If Alexandrov gave public lectures on Ibsen in Chernigov, as he evidently did, he would have had to take positions, at least implicitly, on some political and economic issues that were troublesome in Russia at the time. Perhaps the Whites who took over the city heard that Alexandrov had advanced views unpleasing to them. The consequences for Alexandrov were not serious, however, because General Denikin and his army soon retreated from Chernigov, and Alexandrov was again free.

Soon after this event Alexandrov returned to mathematics; he may have concluded that the theatrical arts were a bit too close to politics. From that time on he almost never expressed political views, certainly not ones that might get him in trouble. He came back to Moscow and Lusitania, where he said he was welcomed as a "prodigal son," and prepared for his master's examinations under the supervision of Egorov, although his official mentor was still Luzin. Alexandrov described Egorov as "the head of the whole of Moscow mathematics."

At this time Alexandrov, now 24 years old, became an affectionate friend of Ekaterina Romanovna Eiges, the sister of his former gymnasium mathematics teacher. Eiges, however, at first did not give Alexandrov her undivided attention because she was romantically attracted to, and meeting with, Sergei Esenin, the famous Russian lyrical poet. (Esenin repeatedly fell in love with both women and men and had five marriages in twelve years, the third of which was to the American dancer Isadora Duncan; he also wrote love letters to his male friend and fellow poet Nikolai Kliuev.) Gradually, however, Alexandrov won Eiges over, probably because for a short time he was more concentrated on her pursuit than Esenin was. By the early spring of 1921, Alexandrov and Eiges were planning marriage. But the situation became more complex because Alexandrov was developing a growing affection for a fellow male mathematics student, Pavel Samuilovich Uryson. It was a triangle of considerable drama.

On March 30 and 31, 1921, two important events occurred in these entangled lives. On the evening of March 30 Alexandrov and Uryson, both passionate about music as well as mathematics, went to a Beethoven concert held in the Bolshoi Theater. After the concert Uryson walked with Alexandrov back to the apartment near the Moscow Conservatory where Alexandrov was living with his musician brother. Arriving there while talking animatedly, the two decided that they did not yet want to part, so they turned and walked to Uryson's apartment near what is presently Maiakovsky Square (about two miles away). Arriving at the new destination, they made the same decision as before, and started back toward Alexandrov's apartment. They kept this up all night, oscillating back and forth, talking the whole time. During the conversation they agreed that they had cemented an intimate friendship which they were sure would continue. They did not leave each other until five o'clock on the morning of March 31. Later both of them referred to that evening as one of the most important of their lives.

On that very same day Alexandrov married Ekaterina Eiges, having had no or very little sleep. The marriage lasted only a few days. Alexandrov later remarked, "Any marriage would have been a mistake for me." He returned to Uryson. It seems clear that this was the moment when Alexandrov decided that he was homosexual and would never marry, at least not successfully.

Alexandrov and Uryson were both preparing for their master's examinations by Egorov, and they often met and "examined" each other in rehearsals for the real event. Gradually their involvement was noticed by other members of Lusitania. Since Alexandrov and Uryson had the same first two initials, P.S., the other Lusitanians often referred to them as "the two P.S.es," or, in the Russian plural, "PSy." The two sometimes referred to each other that way as well. Some of the Lusitanians knew English, and they knew that the term "PSy" could be given a sexual connotation.

Uryson and Alexandrov began spending long times together in

a dacha outside Moscow on the Kliaz'ma River, where they could swim. As Alexandrov reported,

> As soon as we woke up we used to go to the river, which was literally a few steps away from our house. We took with us a large amount of black bread and lightly salted butter. . . . On this food we lasted until about 3 or 4 pm, spending all this time on bathing and on mathematical work, which consisted of the speculations of each us separately, and conversations between us (that is, joint mathematical thinking).

In this way, during the summers of 1921 and 1922, Alexandrov and Uryson did work of fundamental importance in topology. (Alexandrov spoke of the summer of 1922 as a time of "exceptional uplift"; he remarked that the two of them "did mathematics with delight and excitement.") Their work aroused the attention and interest of mathematicians in Germany, Holland, and France, including Emmy Noether, Richard Courant, David Hilbert, Felix Hausdorff, L. E. J. Brouwer, Émile Borel, and Henri Lebesgue.

The two mathematicians continued their work together in the winter months. They established a ritual of going in the evenings to musical concerts and theater events, and afterwards walking the streets of Moscow and talking, as they had first done on that evening of March 30–31, 1921, the memory of which was important to them. Even in winter they did not entirely give up their bathing rituals. Once in December they went swimming together in the Moscow River at 2:00 A.M. in the middle of a snowstorm, breaking through the ice to get to the water.

In the summers of 1923 and 1924 Alexandrov and Uryson traveled together to Germany, Holland, and France where they were welcomed by their foreign colleagues, who already knew of their achievements. In addition to collaborating with fellow mathematicians, the two greatly enjoyed their outings together in Europe—hiking in Norway, swimming across the Rhine in Germany, and climbing onto the roof of Notre Dame in Paris to look at the famous gargoyles. Alexandrov subsequently wrote of the night they stayed in a garret near

Pavel Alexandrov, L. E. J. Brouwer, and Pavel Uryson (left to right), in the garden of Brouwer's home near Amsterdam, 1924.

the Sorbonne, "I have forever kept in my memory that first and last evening I spent there with Pavel Uryson."

On August 9, 1924, the two companions took a train from Paris to Brittany and proceeded to the westernmost point of Finistère, the Pointe du Raz. On this wild and rocky promontory, surrounded by crashing waves, Alexandrov and Uryson hiked and discussed mathematics. Subsequently they went to the small fishing village of Batz-sur-Mer on the coast, where they stayed for several weeks. At first they lived in a small hotel, the Val Renaud, but then moved to a one-room house that was so close to the ocean that the spray from the waves would occasionally come through their window.

Alexandrov described their routine:

> In Batz we walked by the sea, selecting the wildest parts of its stony shore; we swam endlessly and besides we did mathematics. It was there that Uryson wrote his famous paper on

{ 176 }

countable connected Hausdorff spaces, containing many new ideas. I wrote my paper on the topological definition of an n-dimensional sphere.

Almost unnoticed by the two men, the seas were increasing in intensity. Alexandrov remarked that their swims were "getting more interesting." On August 17, at about five in the evening, they went out for another swim. Alexandrov later observed, "When we got into the water, a kind of uneasiness rose up within us; I not only felt it myself, but I also saw it clearly in Pavel. If only I had said, 'Maybe we shouldn't swim today?' But I said nothing."[7]

The seas had risen so much, in fact, that people from the village had gone to this remote spot to watch the waves. They were astounded when they saw the two men plunge into the water and swim for the open ocean. Suddenly a great wave picked them up and threw them back toward the shore. Alexandrov was thrown between two large rocks onto a safe stretch of sand, but Uryson was catapulted directly onto the rocks. Alexandrov soon saw him passively rolling in the waves against the rocks, lifeless. He swam to Uryson, put his arm around him, and tried to paddle back to shore. One of the Frenchmen on the beach threw him a rope and pulled the two men in.

It turned out that one of the men on the beach was a physician on vacation, a Dr. Machefer from Nantes, who gave Uryson artificial respiration and worked over him for a long time. When Alexandrov finally asked him what the situation was, the doctor replied, "Que voulez-vous que je fasse avec un cadavre?" ("What do you wish that I should do with a body?") Alexandrov was traumatized, but later, always the analyst, he remarked that "the word 'fasse' is the present subjunctive form of the verb 'faire' that our French teacher at my school had often asked us for."

Uryson was buried in the presence of a rabbi in the cemetery of Batz-sur-Mer, in a spot where the ocean waves can be heard. His grave, photographed recently by Jean-Michel Kantor, is shown here. It is maintained to this day by the Russian embassy in Paris, which

Grave of Uryson (Urysohn) at Batz-sur-Mer, France.

periodically sends a junior staff member to sweep off the sand and litter that may have accumulated on it, and to place fresh flowers on the stone.

Alexandrov was heartbroken. For a long time, he kept a photograph of Uryson on his desk. In later years he returned to Batz-sur-Mer several times, sometimes by himself, sometimes with others. In the summer of 1925 he spent August there with Uryson's father, who every day between five and six in the evening (the time of the drowning) went to the spot where his son had died. In the meantime, Alexandrov tried to overcome his sadness by engaging in intense mathematical work; he produced a new paper on topology which he sent off to the Dutch mathematician L. E. J. Brouwer, who recommended

Lusitania and After

Pavel Alexandrov.

it for publication in a leading mathematics journal, *Mathematische Annalen*. In October 1925 Alexandrov returned to Batz-sur-Mer with Brouwer, and the two worked together on Uryson's latest papers, still unpublished. Brouwer invited Alexandrov to lecture at the University of Amsterdam in the winter of 1925–26, an offer that Alexandrov accepted, observing that "emotionally I was still completely under the shadow of the heavy loss that I had suffered a year before." In Amsterdam Alexandrov devoted himself to teaching and to regular attendance at musical concerts. In the spring he moved to Göttingen in Germany, where he continued his teaching and musical activities. He subsequently returned to Moscow, and there he gradually became closer to Andrei Kolmogorov, whom he had first met in 1922. Kolmogorov was seven years his junior, and in the first years of their relationship always assumed a subsidiary role. But Kolmogorov would soon become even more famous in the mathematical world than Alexandrov.

Andrei Nikolaievich Kolmogorov (1903–1987) was also a Lusitanian, although he joined the group considerably later than Alexandrov. The circumstances of his birth were unusual: he was born in Tambov to parents who were not married and who played no role in his upbringing. His mother tragically died at his birth; his father, Nikolai Kataev, was an agronomist who soon left Russia only to return shortly after the Revolution to die in fighting. Andrei was raised by his mother's sister, Vera Yakovlena, and he took the family name of his maternal grandfather, a nobleman named Yakov Stepanovich Kolmogorov who had an estate near Yaroslavl. There his aunts ran an excellent small school which Andrei attended. Later, at the age of seven or eight, he moved to Moscow, where he attended a very good private school known as "Repman's gymnasium." Andrei graduated from this school at age 16 and entered Moscow University. Although he was obviously gifted in mathematics, he was also interested in Russian history, especially the history of Russian architecture.

At Moscow University Kolmogorov soon fell under the influence of Nikolai Luzin, Lusitania, and one of Luzin's students, an original member of Lusitania named V. Stepanov. Kolmogorov and Stepanov began to work on trigonometric Fourier series, the same topic that had been at the origin of Cantor's set theory and a subject that Luzin had also studied at the beginning of his work. In 1922, at the age of 19, Kolmogorov published an article entitled "A Fourier-Lebesgue Series Diverging Almost Everywhere," which attracted worldwide attention among mathematicians. During his undergraduate years Kolmogorov published a number of other important papers—eight of them in 1925 alone, the year of his graduation at age 22.

As a young Lusitanian Kolmogorov met Alexandrov, who was at that time closely associated with Uryson. In the years after Uryson's death in 1924 Kolmogorov came to know Alexandrov better, and in 1929 they became intimate friends. In the summer of that year

Lusitania and After

Andrei Kolmogorov.

Kolmogorov and Alexandrov made a long trip down the Volga River in a small rowing boat, continuing by steamship after they reached the city of Samara. They remained on the steamer to the outlet of the Volga at Astrakhan, and then traveled on another ship down the Caspian Sea to the city of Baku, presently the capital of Azerbaijan. The two men continued their voyage on foot and by bus to the large Armenian Lake Sevan, where they stayed for a month on an island in an ancient monastery. Here they swam several times a day in the lake while also pursuing their separate work in mathematics: Alexandrov worked on the topology book which he coauthored with Hopf, while Kolmogorov devoted himself to Markov processes with continuous states and continuous time. Kolmogorov's results from his work by the lake were published in 1931 and marked the beginning of diffusion theory.

Before long Alexandrov and Kolmogorov had established swimming as their favorite recreation, and they continued to enjoy it in the following decades. The next year the two went to France and Batz-sur-Mer, swimming in the same spots along the rocky seashore where Alexandrov and Uryson had once swum together. Back in Moscow, Kolmogorov and Alexandrov began spending long periods in country houses, especially on the Kliaz'ma River, where they could swim. Eventually the two of them purchased a dacha on the river at a spot called "Komarovka," and this house became a favorite meeting spot for some of the world's great mathematicians. The oldest part of the house was built in the 1820s and had been on the estate of a distinguished noble family named Naryshkin; later the house had belonged to a sister of the famous theater director Konstantin Stanislavsky. Although private landowning had been abolished by the time Alexandrov and Kolmogorov bought the house, people were nonetheless permitted to buy vacation dachas on government land. The two men established a routine in which they would usually spend four days each week at the dacha and three days in Moscow, either at the university or at the Steklov Institute of the Academy of Sciences.

Two Cuban students who came to Russia to study mathematics in the 1970s visited Kolmogorov and Alexandrov at Komarovka and left an unusually intimate picture of their lives there.[8] According to their account, the two mathematicians in all seasons except winter alternated between swimming and working on mathematics. The Cubans observed that the most mathematically productive moments usually occurred either just before or just after a swim, when the two men would continue to work together naked, "desnudos." Alexandrov did not have good eyesight, and once when the two were swimming in the nude in sight of some neighboring women on the river, a colleague chastised them for their indecency. According to the story, the near-sighted Alexandrov looked right at the neighbors and insisted, "I don't see any women." Because of his poor eyesight, Alexandrov often swam wearing only his glasses.

Kolmogorov and Alexandrov often seemed oblivious both to their neighbors and to the political environment. In 1931 Kolmogorov became a professor of mathematics at Moscow University. He never

Lusitania and After

Pavel Alexandrov swimming.

mentioned the fact that Egorov, an immediate predecessor and his former teacher, had been arrested and sent to prison, where he starved to death. Alexandrov in his memoirs never mentioned the Stalinist oppression under which they were living. In fact, he remarked that "the second half of the 30s went by peacefully in Komarovka."[9] These were years in which hundreds of thousands of people were arrested, including many known to Alexandrov and Kolmogorov. It was also the time, in 1936, when they were among the accusers of their former teacher Nikolai Luzin in an ideological trial.

A particularly emotional moment for Alexandrov and Kolmogorov came in November 1943, in the city of Kazan, where their teacher Egorov had, over a decade earlier, died tragically after being imprisoned by the secret police. During World War II when Hitler's armies approached Moscow, many factories and scientific institutions were moved farther east, to the Urals and beyond, so they would not fall into the hands of the Nazis. Alexandrov's and Kolmogorov's home institution in the Academy of Sciences, the Steklov Institute, was evacuated to Kazan, and they went with it. On November 25 and 26,

1943, Kazan University staged a jubilee celebration of the 150th anniversary of the birth of Nikolai Lobachevsky, one of Russia's most noted mathematicians, who spent his entire scientific life in Kazan. The archives of Kazan University still contain the program of that celebration, and it shows that both Alexandrov and Kolmogorov participated.[10] Alexandrov gave a talk entitled "Lobachevsky and Russian Science," while Kolmogorov spoke on "Lobachevsky, his Significance and Influence on World Science."

As a part of the celebration the mathematicians present at the conference paid a visit to Lobachevsky's grave in Arskoe Cemetery, located only a few feet from Egorov's unmarked grave. The group accompanying Alexandrov and Kolmogorov included Nikolai Chebotaryov, who had illegally arranged for Egorov's body to be placed there, and Chebotaryov's student, V. V. Morozov, who would, after Stalin's death in 1953, place a monument on Egorov's grave at his own expense. Despite the fact that Egorov's grave was still unmarked in 1943, all the mathematicians knew where it was.

As these men stood before Lobachevsky's tomb, with Egorov's grave immediately behind them, the thoughts that were running through their heads must have been unsettling and chilling. Chebotaryov knew that the reason he was living in Kazan rather than Moscow was that he had lost his job after protesting the dismissal of Egorov from the Civil Engineering Institute for "mixing mathematics and religion." Alexandrov and Kolmogorov lacked that bravery. In 1943 Alexandrov was the president of the Moscow Mathematical Society, a position that Egorov had held until his death in 1931. Kolmogorov was a professor at Moscow University, where Egorov had earlier held sway. Both had benefited from their former teacher's imprisonment. In Kazan, they both gave speeches praising Lobachevsky and lamenting the fact that he had not, during his lifetime, received the recognition of the mathematical establishment in St. Petersburg. Perhaps it was easier for them to be moralistic about events that had happened a century earlier than it was to acknowledge the events in which they were implicated.

A partial explanation for the moral lapses and silences on ethical

Lusitania and After

Alexandrov and Kolmogorov together.

issues on the part of Alexandrov and Kolmogorov may be found in their own relationship. The Soviet secret police gathered information on all prominent people, including scholars, noting their sexual and personal habits. If there was something about an individual that could be used against him or her—such as an unsanctioned sexual relationship or a weakness for alcohol—that information was useful to the secret police even if never actually acted upon. The police could gain control over people simply by making known to their victims what they knew about them. The police soon learned of Kolmogorov and Alexandrov's homosexual bond, and they used that knowledge to obtain the behavior that they wished. When the police asked Kolmogorov and Alexandrov to join in attacking Luzin, they did so. When the government asked them to defend the pseudoscientist Trofim Lysenko, they did so, even though Kolmogorov had earlier criticized the biologist. When, after World War II, the police asked that Alexandrov and Kolmogorov write a condemnation of Alexander Solzhenitsyn, calling him a traitor, they published such a joint letter in the Party newspaper *Pravda*. Kolmogorov on several occasions tried to explain his inconsistencies and disloyalties to colleagues, saying, "Sometime I will explain everything to you." Shortly before his death he stated that he would "fear 'them' [the secret police] to his last day."

In the case of Luzin, however, the criticism of Alexandrov and Kolmogorov was more personal. Luzin resented the fact that the two occupied themselves with new topics—topology for Alexandrov, probability theory for Kolmogorov—that he did not work in; furthermore, the first topic had potentially more general developments than descriptive set theory, of which Luzin was the unique master. Luzin once made a very offensive remark about Kolmogorov and Alexandrov. In 1946, on the floor of the Academy of Sciences, Kolmogorov said something about his recent work on topology to Luzin, and the latter replied, "Eto ne topologiia, eto topolozhstvo" ("This is not topology, this is *topolozhstvo*"). Kolmogorov reddened and struck Luzin in the face. The word "topolozhstvo" is an invented term with a very clear meaning. In Russian the word *skotolozhstvo* means "bestiality," while *muzhelozhstvo* means "sodomy." Therefore, "topolozhstvo" was a word contrived by Luzin that might be translated as "topological pederasty." It was an extremely insulting thing for Luzin to say to Kolmogorov, and the latter's anger is understandable.[11]

Readers may wonder why we include such personal details in this discussion of an important chapter in the history of mathematics. Often mathematicians, and sometimes historians of mathematics as well, pull a discreet cover over anything that might detract from the heroic stature of their subjects. Religion, sex, political pressure, and personal frailties all seem irrelevant to them in their story of the great figures of the field. Lebesgue's and Borel's petty quarrels in the latter part of their lives, Luzin's rudeness, and Alexandrov's and Kolmogorov's failure to defend their colleagues against oppression are often screened out from the later accounts of the development of their disciplines. Some Russian scientists with whom we have talked in recent years have also told us that "they cannot bear to read about the 'Luzin Affair,'" the moment when several internationally known Russian mathematicians denounced their former teacher for personal and political reasons.

We see this history differently. We are not eager to denigrate anyone's accomplishments, and we have an immense respect for science, the most powerful means of knowing developed by human beings. But we think that our understanding of these episodes and, ultimately, our respect for the actors in them will be all the greater if we include the entire story and not just parts of it.

The story of the genesis and development of the Moscow School of Mathematics is the chronicle of great achievements in the history of the field—a record that truly inspires. That this uplifting history also includes religion, sex, and human frailty should allow us to understand more deeply how science and mathematics develop and how human beings live and work.

9

The Human in Mathematics, Then and Now

WHERE DO mathematical ideas come from? This question is a classical one with many different answers over the centuries, starting with the well-known debate between Plato and Aristotle. Plato assumed the existence of an independent world of "ideas," including mathematical ones, while Aristotle claimed that ideas are extracted from "real" things. Mathematicians usually do mathematics without thinking about the origins of their thoughts. But in the exceptional period at the beginning of the twentieth century that we have been examining, there occurred a mixing of mathematical, philosophical, and religious ideas which not only provided an opportunity for questions on the foundations of mathematics to be debated, but sometimes even forced such debate.

Some broad insights can be obtained from our account of French and Russian mathematicians because it concerns central questions in mathematics—its origins as well as the nature and definitions of numbers, space, infinity, and the continuum. The story took place at the time when a *lingua franca*, set theory, was becoming a universal language for mathematicians, but when the limits of the mathematical universe were not yet clarified.

This story can be viewed as a historical experiment in real time which we have reconstructed: If a certain theory has been developed in one country (set theory in Germany) and then is propagated in two others with distinctly different cultural traditions (France and Russia), will it be elaborated in different ways? The answer, at least in this particular case, seems to be yes. At first it was French mathematicians (Borel, Baire, and Lebesgue) who led in the development of Cantor's set theory and did groundbreaking work. Then, under the influence of their ultra-rationalistic traditions, they lost their nerve. The Russian mathematicians (Egorov and Luzin) at first studied at the feet of the French, and then, influenced by their own philosophical and religious traditions, pushed forward toward the creation of descriptive set theory, working with exceptionally talented students.

One should not overstate these cultural differences in mathematics, however. In the early and final stages of our story the attitudes of French and Russian mathematicians working in set theory were rather similar. At the end of the nineteenth century and the first years of the twentieth, the initial reactions by mathematicians and philosophers in both France and Russia were a mixture of acceptance and rejection. Then came the first use of the mathematics of infinities by Borel, and especially Baire, in France, and by Egorov in Russia. At this early stage there were individual, personal differences in approach, but the global context (that is, the distinct cultural differences between the two countries) did not seem to dominate. Then, after Lebesgue's remarkable work, the French mathematicians seemed more and more reluctant to pursue the use of transfinites, and the torch was picked up by the Russians and carried forward to result in the birth of a new discipline, descriptive set theory. In this period (roughly 1910–1925), the cultural, ideological, and religious settings in the two countries played important roles. Later, after the end of Lusitania in the mid- to late twenties and, especially, after 1930 with the publication of Luzin's work in French, the two sides once again became more similar, agreeing on basic definitions and principles.

It is not necessary to resolve the ultimate problems in the philosophy of mathematics in order to see that Name Worshipping—a religious viewpoint regarded as heresy by the Russian Orthodox Church and condemned by the Communist Party as a reactionary cult— influenced the emergence of a new movement in modern mathematics. In contrast to the French leaders in set theory, the Russians were much bolder in embracing such concepts as non-denumerable transfinite numbers. While the French were constrained by their rationalism, the Russians were energized by their mystical faith. Just as the Russian Name Worshippers could "name God," they could also "name infinities," and they saw a strong analogy in the ways in which both operations were accomplished. A comparison of the predominant French and Russian attitudes toward set theory illustrates an interesting aspect of science: if science becomes too cut-and-dried, too rationalistic, this can slow down its adherents, impeding imaginative leaps.

A contrast between the cold logic of the French and the spirituality of the Russians is not new in cultural history. Leo Tolstoy in *War and Peace* compared Napoleon's Cartesian logic in his assault on Russia with his opponent Kutuzov's emotional religiosity. The novelist describes the Russian general Kutuzov, after the critical battle of Borodino, kneeling in gratitude before a holy icon in a church procession while Napoleon is rationalizing his "miscalculation." Thus Tolstoy saw Borodino as a victory of Russian spirit over French rationalism.

Another example, this one internal to mathematics and closer to the context of this book, is the still inadequately explored contrast between cold logic and high spirituality found in the Greek and Indian mathematical traditions. In the Greek Euclidean tradition there is no general theory of irrational numbers, whereas the Indian Brahmans in the twelfth century included the square roots (karanis) as well as negative and other "numbers" among the divine numbers.[1]

Naming, Name Worshipping, and Mathematics

The idea that a "name" has more in itself than the mere word assigned is very old and goes back at least to Plato's *Cratylus*; the con-

The Human in Mathematics

cept has reappeared many times in succeeding centuries. After all, *logos* is a central concept in western culture.[2] In Russia this idea merged with another mystic tradition and became the belief that by naming God and Christ, and also by praying through different techniques, the believer could attain a union with God, or get as close to the divine as is humanly possible. The modern theory of functions initiated by Baire and Lebesgue after the introduction of set theory led Lebesgue to inquire into a precise extension of the notion of functions, extending the explicit analytic expressions (polynomial, trigonometric) of earlier mathematics, but ones that could still be described or named *(nommées)*. In doing so, Lebesgue and the French school were asking questions that would find a satisfactory framework only twenty years later with the theory of recursivity.[3] But by putting such a strong emphasis on "naming," Lebesgue stimulated Russian mathematicians with an awareness of the religious tradition of Name Worshipping to consider the analogous question when they discovered a new hierarchy of subsets of the continuum that emerged after 1916.

In concluding that mysticism helped Russian mathematicians in the development of descriptive set theory, we have had to overcome our own natural predispositions. Both of us are secular in our outlooks—far from being Name Worshippers ourselves. We did not start out writing this book in order to come down on the side of religion in the infamous science-religion debates that have occupied so large a place in recent public discussions. And although during the writing we have acquired a deeper appreciation of the role that religion (and religious heresy) can play in the development of valuable ideas, we have not changed our views. We trust rational thought more than mystical inspiration.

If we had been participants in the debates described in this book, we would have emphasized that "naming" cannot be equated with "creating." We could name all sorts of mythical characters and subjects which do not and never will exist. But naming mythical characters and subjects is not parallel to the situation of mathematicians who named new infinities. When we name a mythical character, we

{ 191 }

know at the time of naming that it does not exist. The mathematicians we have studied here hoped the new infinities they had just named did exist—although some of their critics had doubts, and sometimes even the original "namers" were a bit nervous about what they were doing.

The Russians who developed descriptive set theory and assigned new names to subsets of the continuum posed the possibility of the existence of new entities in the mathematical universe, and they went on to provide a program for future research which resulted in substantial agreement of mathematicians all over the world about the new entities. That achievement might have occurred without the inspiration of a religious heresy, but, as researchers loyal to the historical record, we maintain that the way it actually occurred was within a context of mystical, Name Worshipping stimulation.

We realize, of course, that intellectual causation of the kind we are discussing here can never be proved. (David Hume doubted that any kind of causation actually exists, even in the physical sciences.) In intellectual history, instead of demonstrating causation, all we can do is to build a strong circumstantial case for the filiation of ideas. We believe we have built such a case for the thesis that Name Worshipping influenced the work of the founders of the Moscow School of Mathematics, Dmitri Egorov and Nikolai Luzin.

The Moscow School of Mathematics was one of the most powerful movements in the field of mathematics—a movement that is still alive today, even though many of its recent products have emigrated to other countries such as the United States, Israel, France, and Germany. Many of these current descendants of the Moscow School of Mathematics know nothing about the Name Worshipping proclivities of the founders of the field.

Name Worshipping as an influence in mathematics has now largely disappeared, but interest in the general subject of Name Worshipping, particularly its historical, philosophical, and religious dimensions, is once again growing. In the last ten years approximately fifteen books have been published in the Russian language on Name Worshipping.[4] And even though Lusitania, properly speaking, has not existed for eighty years, something of its unusual spirit may still be alive. Many foreign mathematicians who have visited Moscow and attended mathematics seminars there have been struck by the intellectual (even semi-religious) intensity they found. Nikolai Luzin, as the Russian mathematician V. M. Tikhomirov recently observed, "changed the style of Moscow mathematical life" to that of a "passionate love of, and selfless interest in mathematics."[5]

Was there more to this style than devotion to mathematics? Luzin not only possessed a mystical view of the universe that infused his teaching; he also worked with students in an entirely new way. Tikhomirov went on to say, "Luzin would start from the outset by posing to his students, who were hardly out of high school, problems of the highest level, problems that stymied the most eminent scholars." Furthermore, Luzin believed that mathematics has no boundaries, that it touches upon all of life. Tikhomirov observed that when Luzin's students had tea with him in his apartment, "the conversation touched upon the most varied cultural topics." Lusitania had what Kolmogorov called "a common heartbeat," a shared intensity of view toward not just mathematics but all of culture. Many of the members of Lusitania were also passionate about music, literature, architecture, and art.

In recent years many mathematicians educated in Russia have emigrated to other countries, where they now live, teach, and do research in places like Paris, Berlin, both Cambridges, Tel Aviv, Princeton, Berkeley, Ann Arbor, Minneapolis, and many other locations. When we have asked non-Russian members of mathematics departments in these cities if the arrival of the Russians had an effect on their mathematical lives, we have heard this comment frequently: "They brought a different mathematical style."

Describing this style is not so easy, but an example is provided by Jean-Michel Kantor's experience in the late 1970s when he was beginning his education in mathematical research in France while at the same time making regular visits to seminars at Moscow University. Some of his young mathematician friends in Russia were not only doing very good new mathematics but also propagating mathematics to Russian schoolchildren through "Olympiads" and in articles published in the extraordinary magazine *Quantum*, created by Kolmogorov. We would not go so far as to say that the passion with which these young Russian students thought day and night about mathematical questions was unique; the same passion could be found in the corridors of mathematics departments in Cambridge, Bonn, Berkeley, Paris, and elsewhere. One characteristic of the Russian approach, however, stood out—the conviction of the best Russian teachers of mathematics that the most fruitful attack on problems was direct and straightforward, without any preliminary, long, heavy readings. In other words, *start from scratch*. By doing so, one got an almost physical feeling of being directly in contact with mathematical objects and experienced the sensual pleasure of having to fight intellectually with one's bare hands. One of the great mathematicians of the time, Israel Moissevich Gelfand, would tell his young students, "We should study this topic before it has been tainted by handling" ("Etu temu nado izuchat' poka ne zakhvatali").

We saw the birth of this style in Chapter 6 when we described how Luzin just before World War I transformed mathematics teaching at Moscow University; he involved his students in intellectual interchanges so absorbing that when the seminar formally ended, the students would follow him out of the building and then all the way to Luzin's apartment, where the debates continued past midnight. Thus, something of Lusitania lives on today in the corridors of universities and in the homes of mathematicians not only in Russia, but in many other countries. Though diluted and transformed, the spirit of Lusitania is still alive. Russia has influenced the world of mathematics.

Of the three main Russian figures in our story—Egorov, Florensky, and Luzin—the first two were definitely, by their own admission and assertion, Name Worshippers. Luzin was never so public about his religious views, but we know from his letters and his readings (especially in the period 1908–1910) that he was deeply motivated by religious mysticism and its link to mental creativity. He was somewhat secretive about his religious commitments—with good reason in the Soviet period—but his religious beliefs were obviously strong and close to those of his Name Worshipping friends. Luzin, too, was obsessed with "naming," as his archival mathematical notes demonstrate (see the Appendix). And after World War II, when Stalin loosened the controls over religion in order to elicit the patriotic support of the church in the struggle, we are told that Luzin resumed going to church, and continued doing so until his death in 1950.

The Human Side of the Story

We have worked to identify intellectual motivations in our story, but our interests include larger considerations. Like everyone else, we are also emotional human beings affected by the fates of people around us and, in this instance, by the lives of the subjects of our scholarly inquiry. Is there anyone—however far he or she may be from Name Worshipping mysticism—who would not be deeply moved by the stories of Dmitri Egorov, Pavel Florensky, and Nikolai Luzin? All three were cruelly persecuted because of their "anti-Soviet" personal viewpoints and suffered almost unspeakable hardship. Egorov and Florensky died in detention (by starvation and execution, respectively), and Luzin was publicly humiliated and barely escaped a similar destiny. (And these three are far from the only ones in our story to suffer in such a way; think of the death of Luzin's lover, Nina Bari, who threw herself under the wheels of a subway train in Moscow, or the suicide of Luzin's student Shnirel'man in the kitchen of his Moscow apartment.)

Are these people "heroes"? In our skeptical age, most people are reluctant to use the term. Certainly Egorov and Florensky were extremely brave men. When Egorov was accused publicly of being a "wrecker," a person trying to sabotage the Soviet cause, he replied sharply that the real "wreckers" were those who insisted on one ideological viewpoint, denying the diversity that leads to creativity. In speaking in this way, he was sentencing himself. Florensky insisted on wearing his priest's robe while giving papers at Soviet scientific meetings, again attracting the attention of the secret police. As the intensity of Soviet oppression increased, very few people dared to defend themselves openly in the ways in which Egorov and Florensky did. Luzin chose, like most others, to be silent about his beliefs and preferences, although his silence did not, in the end, keep him out of trouble. But that silence may have played a role in the fact that he, alone of the three, died a natural death. He was a less brave man, but a more creative one, at least in mathematics.

All three men, like all of us, displayed weaknesses. Egorov showed poor judgment when he refused to help students in his mathematics classes in the 1920s who were young Communists. Florensky was probably, in some of his writings, anti-Semitic. Luzin was somewhat unstable psychologically and perhaps a bit ungenerous to his brightest students. He also gratuitously insulted his homosexual former student and later colleague Kolmogorov, creating an enmity that would have consequences for him later.

Still, we admire these men and respect them for what they did. If we had to choose an individual in our story whom we admire the most, however, it would be an entirely different person, one who played a relatively minor role: Nikolai Chebotaryov. The reason is that he was one of the few people in this story who displayed tolerance for persons with views different from his own—a characteristic needed in all societies, and perhaps particularly in Russia. It was Chebotaryov—mathematician, young veteran of the Red Army, and loyal Soviet citizen—who resigned his hard-won teaching position at an engineering institute in Moscow when he learned that the man he replaced, Dmitri Egorov, had been fired because of his "mixing of mathematics and mysticism." As a result of his principled action, based only on human sympathy rather than on ideological agreement, Chebotaryov was forced to take a teaching position in far-off Kazan. And years later, when Chebotaryov and his wife, Maria Smirnitskaia, were once again presented with a moral challenge by Egorov, they did not shrink from it. On the contrary, they did everything in their power to try to save Egorov's life after he was sent in exile to Kazan. After failing in this task, they arranged for Egorov to have a decent burial and funeral when others feared to defend a person accused of treason against the Soviet state. For this action Nikolai Chebotaryov was, once again, punished; he was blackballed by his superiors from election to one of the highest (and financially most remunerative) honors in Soviet society—full membership in the Academy of Sciences. As far as we have been able to establish, Chebotaryov was not religious and shared none of Egorov's mystical beliefs. The actions of Nikolai Chebotaryov and Maria Smirnitskaia were governed by a sense of fairness, not by ideological or religious commitment.

Science and Religion

In this spirit, and within the context of this book, let us examine briefly the issue of science and religion. The story we have examined here demonstrates what historians of science have long known—that religious belief can, at least in some instances, facilitate scientific creativity. Anyone familiar with the work and beliefs of Isaac Newton or Blaise Pascal, to name only two individuals, can find such evidence. Of course, religion can conflict with science as well, more dramatically and even disastrously, as the controversies surrounding Galileo and Darwin demonstrate. In the story we have told here the main theme has been the "supporting" rather than the "conflicting" one, since we have maintained that Name Worshipping facilitated mathematical creativity in this period. But we should point out that there is also an episode in this book where we described atheism as helping mathematical creativity: the development by A. A. Markov of Markov Chains, inspired by his opposition to his colleague P. A. Nekrasov's attempts to justify a theological concept of free will by a "pairwise independent" interpretation of the Law of Large Numbers, as described in Chapter 4. Intellectually, we see as much strength in Markov's views as we do in those of the Name Worshippers. In these two different cases, Markov and Egorov-Luzin, two contrasting viewpoints—atheism and religion—both helped motivate mathematicians to creative achievements.

In our opinion, therefore, it is simplistic to insist on either an inherently "conflictual" or an inherently "harmonious" relationship between science and religion. One must look at the contexts and details of individual cases, without prejudging the issue.

What Next?

Before envisioning the future, let us first emphasize that both mathematical and philosophical developments of the problems we described are still influencing thinkers. Descriptive set theory (DST) has lived on and influenced the development of mathematics in the second part of the twentieth century. Two developments have occurred since the end of the period we have been considering. First, there was a confluence of mathematics with logic, through the work of Church, Kleene, Turing, and others on computability and recursion theory, and also through the first results of Kurt Gödel, who confirmed in 1938 that the remarkable intuitions of Luzin were correct and showed that the difficult questions raised by the study of analytic and projective sets could not be solved in the framework of classical set theory (Zermelo-Fraenkel axioms). Many other remarkable developments have occurred since then, even recently, including a radically new approach to the Continuum Hypothesis.[6]

Second, DST also had an impact on modern probability theory through the theory of capacities developed in the 1950s by Gustave Choquet, a student of Arnaud Denjoy, after he spent two years in Poland where he studied with the Polish school of Sierpinski. Choquet's work permitted the subtle analysis of trajectories of Brownian movements and other phenomena in probability theory.[7]

The theological and philosophical questions debated here have been pursued as key issues throughout the last century. The religious heresy of Name Worshipping was informed by the recent revival of theological debates about the fundamental alliance of early Christianity with the Hellenistic tradition.[8] The ontological debate about realism and the existence of mathematical objects led to an impressive number of articles, books, and debates, among them those by

The Human in Mathematics

Quine and Dummett in the English-speaking world and by Desanti and Bouveresse on the French side.[9]

Turning now to the future of mathematics, what does it hold, in view of the events we have described? Georg Cantor opened a Pandora's box in mathematics when he announced that "the essence of mathematics is freedom." Today that box is still not closed. Whose freedom? Luzin and Brouwer gave their own answers—Luzin by giving mathematics a psychological and metaphysical dimension, Brouwer by trying to reconstruct all of mathematics on an intuitionistic basis. Both approaches later suffered reverses, with the descriptive set theory of Luzin being crowded to the edge of the stage by Alexandrov's topology (as well as by the brutality of Luzin's trial), the gradual defeat of the school of Brouwer by the axiomatics of Hilbert and, later, by Bourbaki's domination.

The relevance of a human direct involvement in mathematics appeared in various cases in the nineteenth century, for example, in the work of George Boole, who wanted to "investigate the fundamental laws of those operations of the mind by which reasoning is performed," and in this way founded modern logic. Such involvement received a strong impulse in the twentieth century with the crisis in set theory (see the discussion among the French mathematicians about the "Axiom of Choice" and what choosing means). The issue grew in importance again when the first computer-aided proofs appeared.[10] More recently, the issue gained importance with the maturation of the immense personality and work of the mathematician Alexander Grothendieck in France. Grothendieck gave a more radical meaning to Cantor's freedom, not only by pushing into unknown territories the notions of space and including even logic in his geometry, but also by making the "continuum/discrete" aporia explode. He involved his personality as a mathematician in his work by intricately mixing conjectures with an immense program of research.

In connection with our topic, it is remarkable to see the similarities (probably not due to the common Russian origins of Grothendieck and Luzin) between the Luzin tradition and the description by

Grothendieck of his creative process—the birth of his "messenger dreams" and "visions," as he calls them in his deep and poetic autobiographical work.[11] Furthermore, Grothendieck, like Luzin, placed a heavy emphasis on "naming," seeing it as a way to grasp objects even before they have been understood.[12]

The oppositions of the continuum and the discrete, of rationalistic and intuitionist approaches, which have been present through the whole history of science and culture are still debated in recent intellectual discussions. It has even been suggested that the personal flow of time for the individual is not of a continuous but of a discrete nature, and, moreover, that there could be physiological roots for the dilemma of the continuum versus the discrete.[13]

In the history of mathematics, from the time of Pythagoras to the present, there have been periods of waxing and waning of the elements of rationalism and mysticism, or, perhaps more accurately, rationalism and subjectivism. In the first years of the twentieth century, at the time of deep discussions of the new set theory, subjective elements were particularly important, as mathematicians wrestled with foundational questions that involved philosophical and religious considerations. Subsequently, in the 1930s, 40s, and 50s, subjectivism retreated under the assault of ideas coming from Bertrand Russell, Gottlob Frege, the Hilbert school of axiomatics, and later, the Bourbaki school in France.

Will this period be followed by a new wave of subjectivism? Will the history of mathematics from the beginning of the last century through the present one look like an arc, starting in a period of subjectivism, then tracing the ascendance of analytical rationalism, and then, in recent years, returning to greater subjectivism? People favoring this view could cite the work and opinions of Grothendieck, still alive, but already a legendary mathematician as well as the author of a long and fascinating autobiography with strong mystical dimensions. Is Grothendieck a harbinger of a new era, or is he an exception who falls outside general patterns, a true and unique genius?

The Human in Mathematics

Maybe Grothendieck, with his idiosyncratic mystical views tightly interwoven with meta-mathematics, is indeed a prelude of the future development of mathematics. Grothendieck has insisted, however, that mathematicians do not need religion.[14] Our belief, as we have shown here, is that sometimes it can help.

APPENDIX

NOTES

ACKNOWLEDGMENTS

INDEX

APPENDIX

Luzin's Personal Archives

Introduction

The archives of Nikolai Luzin were partially investigated by Roger Cooke during a period of three months in the winter of 1988–89 at the Academy of Sciences in Moscow.

In the material that follows, Luzin's texts are printed in boldface, and "fr" means that the text is in French. Numbers in brackets refer to the list of references at the end of the Appendix.

We would like to thank Roger Cooke for kindly transmitting to us his notes, which consist of (sometimes annotated) copies of Luzin's manuscripts, and guiding us into the labyrinth of these truly extraordinary documents. Remarkably, in Luzin's papers, mathematics (such as the solution to Baire's problem in 1914; see [4]) are often interwoven with philosophical and personal comments; as Cooke observed, Luzin's work seems to be permeated with psychological considerations. Until around 1920 Luzin's remarks were often written in pre-revolutionary Russian script (O.R.), but sometimes the play with modern and old characters is subtle. For example, in about 1918, after a long development in new characters on the structure of the set **N** of integers, Luzin suddenly worried about a possible paradox and turned to O.R. to write emotionally:

This is a lesson to you! Do not lose your head, keep calm and patient, with faith in God and prayers for his Mercy, all, I repeat, will work if your goal is good.

And then he went directly back to mathematics.

Thus it is possible to follow the evolution of Luzin's thought, taking into account as he did the threats to his freedom after 1917.

The reader is referred to [3] for a detailed analysis of the mathematics. We are presenting here only some characteristic pieces of the mathematico-philosophical content of the archives. We maintain that the issues of Name and Naming are present throughout these archives (there are many examples), giving a very strong indication that Luzin was aware of similarities with the theological aspects of the process of Naming in *imiaslavie* (Name Worshipping).

A. Cantor's Diagonal Method

Starting early in 1910, while thinking about the famous argument of Cantor, Luzin meditates on what the existence of a real number means. Luzin says that Cantor's argument only shows that the reals are not "effectively enumerable" (meaning defined without the Axiom of Choice). But it still could be the case that the reals are "countable" but not "effectively enumerable" (an expression of Borel). These topics were still discussed later, of course, in connection with recursivity theory, and also with Gödel's and Cohen's results (see [2]).

B. *Nommer c'est avoir individu*

The following selections from the archives are probably from 1915; they concern the status of mathematical objects, whether they are supposed to exist, or defined with or without the Axiom of Choice. The discussion of "The Five Letters," amplified by Zermelo's construction and Richard's paradox, appeared on various occasions in Luzin's manuscripts for nearly thirty years.

Appendix

Luzin starts with comments on the article by Lebesgue [5] on "fonctions nommables"; then he observes:

> It would be very desirable to have a definition of *nommer*, but it seems impossible at present. I propose:
> Nommer c'est avoir individu. This seems a natural definition since the notion of individuum appears to be a rather primitive one. So that it does not need further definitions. But difficulties appear when one looks at examples.

Comment by the authors: Luzin probably means that when one names a thing, he "singles it out." The correct French expression would be "Nommer, c'est avoir à faire avec un individu."

The examples are highly technical, connected with the Continuum Hypothesis and new delicate notions like "ensemble clairsemé" (sparse set), introduced by Arnaud Denjoy, a longtime friend of Luzin (see [4]).

C. Existenz

Struggling for nearly fifty years with the Continuum Hypothesis, Luzin wrote on many occasions about the second class of transfinites $\aleph_1$, by definition (Cantor) the ordered set of all denumerable ordinals. For example in January 1917, responding to Sierpinski's caveat against definitions made with the Axiom of Choice, Luzin wrote:

> Let us concern ourselves with psychology. We, in our mind, consider natural numbers *objectively existing*.
> We, in our mind, consider *the totality of all* natural numbers *objectively existing*. We, finally, consider the *totality of all transfinites* of the II Class *objectively existing*.
> We want the following: having assumed that we face the objectively-existing totality of all natural and transfinite numbers of the II Class, we connect with each of the transfinites of the II Class a definition, a "name"—and more-

Appendix

over uniformly for all those transfinites we are considering. You see, if we are given naturals, we can write each of them in a decimal system, equally measured, with the symbols 0,1,2,3,4,5,6,7,8,9. There is no need for us to (theoretically) write the symbols in definite places defined exactly as a finite number. Practically, we all the time are located within the limits of millions, or $< 10^{10}$, or something like this.

It is necessary to remember that $4^{4^{4^4}}$ is already a kind of unimaginable number, even in a decimal system. One way or another, then, we connect with each natural number a definite representation of it, equally measured, with ten symbols 0,1,2,3,4,5,6,7,8,9. I say that there is no vicious circle here theoretically because we are talking about the places for symbols located in a finite number which we practically have as < 10, we do not worry about this theoretically, considering that this number *cannot be expressed* in decimal symbols. If we did express it in decimals (i.e., a number of decimal places of a given number N), we would come to a smaller number which we could again express in decimals and so forth. We, taking numbers < 10, calculate a number of processes of reduction and so forth. Our thought would begin to wander in some kind of thick forest of reductions, the reduction of reductions, and all this would be some kind of chaos of reductions, to the effort of thought to embrace the given natural in a single perception in decimal symbols. This goes synthetically.

In a similar way in transfinite numbers of the II Class, it is completely justified to seek something in the nature of a decimal system which would permit one to define (to name — "*nommer*" fr) each transfinite of the II Class. There is no need for us to meet here a theoretical vicious circle, as in finite numbers. In natural numbers each natural number we can *nommer* by means of decimal symbols, *exactly to name*,

Appendix

for the perception of the number of decimal symbols themselves, say N1, can be again a reduction.

But in natural numbers such *nommation* (fr) is done.

Comment: The Continuum Hypothesis, the main problem for many years, can be expressed as the study of $\aleph_1$, the number of all denumerable "well-ordered" sets, and its comparison with the cardinal of the continuum. This means describing all such denumerable ordered sets, naming them. This text is a vague and overly optimistic projection of what Luzin was hoping to be able to do in a parallel between finite and transfinite denumerable ordinal numbers (in fact this is only possible for constructible ordinals; see [1]); these remarks of Luzin are a surprising premonition of the work of Church. We know now that a complete description of $\aleph_1$ is hopeless, because of Cohen's results.

> We see that in the absence of an *analytic* rule *(non-auswahlich)* — which is the only thing that could give us confidence in the existence of the second class or any other — the existence of a class becomes mysterious and the problem is actually the question of the validity of this existence and the very *meaning* of this existence. An analysis of the word "existence" would be interesting! Philosophically it denotes *absolute being*. Only I don't know whether that is equivalent to *objective being*. To exist does not mean at all "to be an object of our thought." It is something more, since even a contradiction can be an object of our thought, and it is deprived of existence. Indeed we speak of *objective* existence of the same degree of certainty as the existence of any mathematical object (in the earlier sense), such as a straight line or a circle. There are two types of *existenz: First*, a thing exists because it is *analytically* defined for everyone; here we do not care what particular analytic procedure is used for the definition; all that matters is that the definition be *analytic*.... We require only operations with arbitrariness elim-

{ 209 }

inated. *Second,* a thing exists by virtue of the Axiom of Zermelo, that is, it exists, although it cannot be analytically defined. That is the true meaning of Zermelo's Axiom. It contains the concept of "existenz," and therefore everything reduces to uncovering the content of that concept.

Comment: At this point, Luzin's ideas are still very influenced by Kantian ontology, but his mathematical views seem to be at least tentatively more precise than the French ones of 1905. He makes a clear distinction between construction with or without the Axiom of Choice.

D. Discovery of A-sets (Suslin, Luzin, summer 1916)

The importance of the discovery of Luzin and Suslin in 1916, which marked the start of Lusitania and of descriptive set theory, was well understood from the beginning by Luzin, as is shown by the length and depth of the archives devoted to the topic: the proof of Suslin written with very critical appreciation by Luzin, who was unhappy with Suslin's description (10 pages), followed months later by a new proof by Luzin. Luzin also noted the specificity of the mathematical ideas involved—with Cantorian new notions playing an essential role in definitions and notations, as he writes at a certain point in the mathematical considerations:

> **Everything seems a daydream and game of symbols, which give, however, the most important results.**

And also:

> **Each definition is a piece of secret ripped from Nature by the human spirit. I insist on this: any complicated thing, being illumined by definitions, being laid out in them, being broken up into pieces, will be separated into pieces completely transparent even to a child, excluding foggy and dark parts that our intuition whispers to us while acting, separat-**

ing into logical pieces, then only can we move further, towards new successes due to definitions . . .

A temporary conclusion to this dense material in a few hundred pages of archives can be found in the profound article written by Luzin in 1928 for the International Congress of Mathematicians in Bologna [6]. There, confronting the recent developments of set theory with Brouwer's intuitionism and Hilbert's new views on "metamathematics," Luzin takes a mixed position between "idealists" and "realists"; examining the "enigmatic resistance" of many mathematical problems, he is not far from doubting that some of these problems will ever find a solution when he describes the "fatigue du paradis de Cantor" (the fatigue with Cantor's Paradise). We know now that he was right and prophetic.

References

[1] Choquet, Gustave. "Epistémologie du transfini." *Cahiers du Séminaire d'histoire des mathématiques*, no. 1 (1980), pp. 1–17.

[2] Cohen, Paul. *Set Theory and the Continuum Hypothesis*. New York: W. A. Benjamin, 1966.

[3] Cooke, Roger. "Arkhiv Luzina," *Istoriko-Matematicheskie Issledovania*, 34 (1993), pp. 246–255; review of J. Ferreiros, in *Labyrinth of Thought: A History of Set Theory and Its Role in Modern Mathematics, The Review of Modern Logic*, vol. 9, 1 and 2, 29 (2001–2003), pp. 167–182.

[4] Dugac, Pierre. "Nicolas Lusin: Lettres à Arnaud Denjoy, avec introduction et notes." *Archives Internationales d'histoire des sciences*, 27 (1977–1978), pp. 179–206.

[5] Lebesgue, Henri. "Sur les functions représentables analytiquement." *Journal de mathématiques pures et appliquées* (6), 1 (1905), pp. 139–216.

[6] Luzin, N. N. "Sur les voies de la théorie des ensembles." *Atti del Congresso Internationale dei Mathematici*, vol. I. Bologna: Zanichelli, 1928, pp. 295–299.

NOTES

1. Storming a Monastery

1. For these and other details on the Mt. Athos events, see "Imiaslavie," *Nachala* (no. 1–4) (Moscow, 1996); Episkop Ilarion (Alfeev), *Sviashchennaia taina tserkvi: vvedenie v istoriiu i problematiku imiaslavskikh sporov*, vols. I and II (St. Petersburg: Aleteiia, 2002); *Imiaslavie: sbornik bogoslovsko-publitsisticheskikh statei, dokumentov i kommentariev* (Pskov and Moscow, 2003); A. M. Khitrov and O. L. Solomina, *Zabytye stranitsy Russkogo imiaslaviia* (Moscow: Palomnik, 2001); Dimitrii Leskin, *Spor ob imeni Bozhiem: filosofiia imeni v Rossii v kontektste Afonskikh sobytii 1910-kh gg.* (St. Petersburg: Aleteiia, 2004).
2. J. D. Salinger, *Franny and Zooey* (Boston: Little, Brown, 1961), pp. 36–37.
3. Skhimonakh Ilarion, *Na gorakh Kavkaza*, 3rd ed. (Kiev: Kievskaia Perchersskaia Lavra, 1912), *passim*.
4. *Imiaslavie: Materialy k razresheniiu spora ob Imeni Bozhiem*, Nachalo (1–4, 1995) (Moscow, 1996), p. 23.
5. René Fülöp-Miller, *The Mind and Face of Bolshevism* (London, 1926), p. 258.
6. Leskin, *Spor ob imeni Bozhiem*, p. 147.
7. Ibid., p. 266.
8. See "Bozhestvennaia onomatologiia dekonstruktsii: Zhak Derrida," in Elena Gurko, *Bozhestvennaia onomatologiia: imenovanie Boga i imiaslavii, simvolizme i dekonstruktsii* (Minsk: Ekonompress, 2006), pp. 317–432.

Similar current interest in the ideas of Florensky was shown at the conference "Days of Science" sponsored by the Dynasty Foundation in St. Petersburg, April 21–22, 2008, where Graham and Kantor presented papers. See Elena Kokurina, "Matematiki svobodny, kak i filosofy," *V Mire Nauki* (no. 6, 2008), pp. 12–13.

2. A Crisis in Mathematics

1. Quoted in José Ferreirós, *Labyrinth of Thought: A History of Set Theory and Its Role in Modern Mathematics* (Basel: Birkhäuser Verlag, 1999), p. xi.
2. Hermann Weyl, "Über die neue Grundlagenkrise der Mathematik," *Mathematische Zeitschrift*, vol. 10 (1921), pp. 39–79.
3. Hermann Weyl, *God and the Universe: The Open World* (New Haven: Yale University Press, 1932), p. 8.
4. *Physics* VI: 9, 239 b15.
5. See Bertrand Russell, *The Principles of Mathematics*, vol. 1 (Cambridge: Cambridge University Press, 1903).
6. See Harvey M. Friedman quoted in A. Borovik's blog: *http://www.maths.manchester.ac.uk~avb/micromathematics/2006/09/achilles-tortoise-and-yessenin-volpin.html*, pp. 2–3.
7. G. H. Moore, "Towards a History of Cantor's Continuum Problem," in David E. Rowe and John McCleary, eds., *The History of Modern Mathematics*, vol. I (Boston: Academic Press, 1989), pp. 79–122; Ferreirós, *Labyrinth of Thought*.
8. Quoted in Ferreirós, *Labyrinth of Thought*, p. 265, and letter of November 5, 1882, to Dedekind.
9. G. Cantor, "Über unendliche, lineare Punktmannlichfaltigkeiten," *Mathematische Annalen*, reference to Cantor, *Gesammelte Abhandlungen* (Zermelo, ed.), 1932.
10. Letter from Hermite to G. Mittag-Leffler, December 24, 1880, as quoted in Pierre Dugac, *Histoire de l'analyse*.
11. "Leur tournure philosophique ne sera pas un obstacle pour le traducteur qui connaît Kant." Charles Hermite, "Lettres à G. Mittag-Leffler publiées et annotées par Pierre Dugac," *Cahiers du Séminaire d'Histoire de Mathématiques*, 5 (1980), pp. 49–285.
12. "Ils ont un peu l'air d'une forme sans matière ce qui répugne à l'esprit francais." P. Dugac, ed., "Lettres de Poincaré à Gosta Mittag-Leffler," *Cahiers du Séminaire d'histoire des mathémathiques*, 5 (1984), p. 205.

3. The French Trio

1. "Different infinities" means that there cannot be any one-to-one correspondence between the integers and the continuum.
2. Nicolas Boileau-Despréaux, *Les quatre poétiques d'Aristote, d'Horace, de Vida, de Despréaux*, vol. 1 (Paris: Saillant et Nyon-Desaint, 1771), p. 153.
3. E. Picard, *Une édition nouvelle du discours de la méthode de Descartes* (Paris: Gauthier-Villars, 1934), p. 54.
4. See B. Belhoste, "L'enseignement secondaire français et les sciences au début du XXe siècle: La réforme de 1902 des plans d'études et des programmes," *Revue d'Histoire des Sciences*, 53 (4) (1990), pp. 371–400.
5. The details of Borel's life have been drawn from Pierre Guiraldenq, *Emile Borel, 1871–1956, L'espace et le temps d'une vie sur deux siècles* (Paris: Albert Blanchard, 1999); and Camille Marbo, *À travers deux siècles: Souvenirs et rencontres (1883–1967)* (Paris: Grasset, 1967).
6. Charles Hermite-Stieltjes, *Correspondance* (Paris: Gauthiers-Villars, 1905), p. 318.
7. See Bernard Maurey and Jean-Pierre Tacchi, "La genèse du théorème de recouvrement de Borel," *Revue d'histoire des mathématiques*, 11, no. 2 (2005), pp. 163–204.
8. *Collection de monographies sur la théorie des fonctions* (Paris: Gauthier-Villars, 1950).
9. "Après avoir intrigué pour soi, on intrigue d'abord pour ses enfants, qui sortent du collège, puis pour les gendres, puis pour leurs enfans au berceau; le système des familles naturelles envahit le sanctuaire; et conduit par un bras tout-puissant, il faut qu'un gendre soit bien lourd, pour se laisser devancer à la course par un parvenu non indigène." Quoted by P. Corsi, *Genèse et enjeux du transformisme, 1770–1830* (Paris: Editions du CNRS, 2001), p. 305.
10. Camille Marbo, *À travers deux siècles*.
11. The biographical information on Lebesgue comes from B. Bru and P. Dugac, eds., "Lettres d'Henri Lebesgue à Émile Borel," in Henri Lebesgue, *Les Lendemains de l'intégrale: Lettres à Émile Borel* (Paris: Vuibert, 2004).
12. Quoted by L. Félix in *Message d'un mathématicien, Henri Lebesgue, pour le centenaire de sa naissance* [Texte imprimé]—*introduction et extraits choisis par Lucienne Félix, préface par S. Mandelbrojt* (Paris: A. Blanchard, 1974), p. 124.

13. Lebesgue, "Sur les fonctions représentables analytiquement," *Journal de Mathématiques Pures et Appliquées*, I (1905) p. 205; also in his *Oeuvres scientifiques*, vol. III (Paris, 1972), p. 169.
14. "Je ne sais pas s'il est possible de nommer *une* fonction (italics in the original) non mesurable; je ne sais pas s'il existe des fonctions non mesurables." Lebesgue, *Leçons sur l'intégration et la recherche des fonctions primitives* (Paris: Gauthiers-Villars, 1904), p. 112; also in his *Oeuvres scientifiques*, vol. II (Paris, 1972), p. 128.
15. Borel, "Sur la représentation effective de certaines fonctions discontinues," Note de E. Borel, *C.R.T.*, 137 (1903), pp. 903–905.
16. Lebesgue, *Leçons sur l'intégration;* italics in the original.
17. The information on Baire's life comes from P. Dugac, "Notes et documents sur la vie et l'oeuvre de René Baire," *Archive for the History of the Exact Sciences*, 15 (1976), pp. 297ff.
18. E. Zermelo, "Beweis, dass jede Menge wohlgeordnet warden kann," *Mathematische Annalen*, 59 (1904), pp. 514–516.
19. J. Hadamard, R. Baire, H. Lebesgue, and E. Borel, "Cinq lettres sur la théorie des ensembles," *Bulletin de la Société Mathématique de France*, 23 (1905), pp. 261–173.
20. Hadamard, letter 1 in "Cinq lettres."
21. E. Borel, "Les Probabilités dénombrables et leurs applications arithmétiques," *Rendiconti del Circolo Matematico di Palermo*, 27 (1909), pp. 247–271.
22. Borel's rejection of the Axiom of Choice was spelled out in the Acts of the Fourth International Congress of Mathematicians in Rome, April 6–11, 1908. His opposition to non-denumerable infinities was maintained throughout the prefaces of all four later editions of his *Leçons sur la théorie des fonctions* up to 1950.
23. E. Picard, *La science moderne et son état* (Paris: Flammarion, 1909), extract from chap. 2.
24. Lebesgue, "Sur les fonctions représentables analytiquement," and analysis by Cavaillès in *Remarques sur la formation de la théorie abstraite des ensembles* (Paris: Hermann, 1938).
25. Bertrand stopped his work in pure mathematics and went into probability at the end of his career.
26. Camille Marbo, *À travers deux siècles*, p. 172.
27. In various instances, Borel repeated this statement. For example, see Borel, "La théorie des ensembles et les progrès récents de la théorie des fonctions," *Revue générale des sciences*, 20 (1909), pp. 315–324.

4. The Russian Trio

1. Proceedings of the International Congress of Mathematics, Zurich, 1897. The information on Bugaev's life and work is drawn from S. S. Demidov, "N. V. Bugaev and the origin of the Moscow school of the theory of functions of a real variable" (Russian), *Istoriko-matematicheskie issledovoniia*, 29 (1985), pp. 113–129.
2. Quoted in S. M. Polovinkin, "O studentcheskom matematicheskom kruzhke pri Moskovskom matematicheskom obshchestvo v 1902–1903 gg.," *Istoriko-matematicheskie issledovaniia: sbornik statei*, 30 (Moscow: Nauka, 1986), p. 151.
3. Letter from A. Markov to A. A. Chuprov, November 10, 1910, reprinted in G. P. Basharin, A. N. Langville, and V. A. Naumov, *The Life and Work of A. A. Markov* (Amsterdam: Elsevier, 2004), p. 12.
4. Sources on the life of Egorov include: C. E. Ford, "Dmitrii Egorov: Mathematics and Religion in Moscow," *The Mathematical Intelligencer*, 13(2) (1991), pp. 24–30; P. I. Kuznetsov, "Dmitri Fedorovich Egorov," *Russian Mathematical Surveys*, 26 (1971), pp. 125–164; A. L. Shields, "Luzin and Egorov," *The Mathematical Intelligencer*, 9(4) (1987), pp. 24–27; A. L. Shields, "Luzin and Egorov," Part 2, *The Mathematical Intelligencer*, 11(2) (1989), pp. 5–8; V. Steklov, P. Lazarev, and A. Belopolskii, "Zapiska ob uchenykh trudakh D F Egorova," *Izvestiya Rossiiskoi akademii nauk*, 18 (1924), pp. 445–446.
5. S. S. Demidov, "Professor Moskovskogo universiteta Dmitri Fedorovich Egorov i imeslavie v Rossii v pervoi treti XX stoletiia," *Istoriko-matematicheskie issledovaniia*, 2nd series, no. 4 (39) (Moscow: Ianus-K, 1999), pp. 123–156; see also Charles Ford, "Dmitri Egorov: Mathematics and Religion in Moscow," *The Mathematical Intelligencer*, vol. 13, no. 2 (1991), pp. 24–30.
6. "Stydlivo nizitsia Egorov," Andrey Bely, *The First Encounter*, trans. Gerald Janeček (Princeton: Princeton University Press, 1979), p. 73.
7. S. S. Demidov, "Professor Moskovskogo universiteta Dmitri Fedorovich Egorov i imeslavie v Rossii v pervoi treti XX stoletiia," *Istoriko-matematicheskie issledovaniia*, 2nd series, no. 4 (39) (Moscow: Ianus-K, 1999), p. 138.
8. Sources on Luzin's life include: N. N. Luzin, *Sobranie sochinenii*, Academy of Sciences of the USSR, Moscow, 1953–1959; S. S. Demidov and B. V. Levshin, *Delo akademika Nikolaia Nikolaevicha Luzina* (Saint Pe-

tersburg: RKhGI, 1999); S. S. Demidov, "From the Early History of the Moscow School of Function Theory," *Philosophia Mathematica*, 3 (1988), pp. 29–35; S. S. Demidov, A. N. Parshin, and S. M. Polovinkin, "On the Correspondence of N. N. Luzin with P. Florensky" (Russian), *Istoriko-matematicheskie issledovaniia*, 31 (1989), pp. 116–191; C. E. Ford, "The Influence of P. A. Florensky on N. N. Luzin," *Historia Mathematica*, 25 (1998); C. E. Ford, "Mathematics and Religion in Moscow," *Mathematical Intelligencer*, 13(2) (1991), pp. 24–30; L. V. Keldysh, "The Ideas of N. N. Luzin in Descriptive Set Theory," *Russian Mathematical Surveys*, 29(5) (1974), pp. 179–193; P. I. Kuznetsov, "Nikolai Nikolaevich Luzin," *Russian Mathematical Surveys*, 29(5) (1974), pp. 195–208; M. A. Lavrentev, "Nikolai Nikolaevich Luzin," *Russian Mathematical Surveys*, 29(5) (1974), pp. 173–178; E. R. Phillips, "Nikolai Nikolaevich Luzin and the Moscow School of the Theory of Functions," *Historia Mathematica*, 5 (1978), pp. 275–305; A. L. Shields, "Luzin and Egorov," *Mathematical Intelligencer*, 9(4) (1987), pp. 4–27.

9. Quotations are from S. S. Demidov, A. N. Parshin, and S. M. Polovinkin, "The Correspondence of N. N. Luzin with P. A. Florensky" (Russian), *Istoriko-matematicheskie issledovaniia*, no. 31 (1989), pp. 116–191.

10. Sviashcheniik Pavel Florenskii, *Sviashchennoe Pereimenovanie: Izmenenie imen kak vneshnii znak peremen v religioznom soznanii*, Khram sviatoi muchenitsy Tatiany (Moscow, 2006).

11. Charles E. Ford, "The Influence of P. A. Florensky on N. N. Luzin," *Historia Mathematica*, 25 (1998), p. 334.

12. For some of these details of Luzin's private life, see Ford, "The Influence of P. A. Florensky on N. N. Luzin," pp. 332–339; and M. A. Lavrent'ev, "Nikolai Nikolaevich Luzin," *Russian Mathematical Surveys*, vol. 29, no. 5 (1974), pp. 173–178.

13. Sources on the life of Florensky include: A. P Shikman, Deiateli otechestvennoi istorii (Moscow: Ast, 1997); Frank Haney, *Zwischen exakter Wissenschaft und Orthodoxie zur Rationalitätsauffassung Priester Pavel Florenskijs* (Frankfurt on Main: Peter Lang, 2001); Frank Haney et al., eds., *Pavel Florenskij*—Tradition und Moderne: Beiträge zum Internationalen Symposium an der Universität Potsdam, 5 bis 9 April 2000 (Frankfurt on Main: Peter Lang, 2001).

14. Sergei S. Demidov and Charles E. Ford, "On the Road to a Unified World View: Priest Pavel Florensky—Theologian, Philosopher, and Scientist," manuscript, July 2001, p. 1.

15. *Istoriko-matematicheskie issledovaniia*, XXX (1986), p. 160.
16. Ibid.
17. Charles E. Ford, "N. N. Luzin as Seen Through His Correspondence with P. A. Florensky," *Modern Logic* (July-October, 1997), pp. 233–255.
18. We believe that the title of the essay was as given in our text, although we cannot be certain because we have not actually seen this essay. Its existence was reported as "Ob elementakh alpha-irichnoi sistemi schisleniia," in *Istoriko-matematicheskie issledovaniia*, 31 (1989), p. 134, but there the title was given as "alpha number system," not "aleph number system." We think this is almost certainly either a typographical or a translation error, and that "aleph number system" was intended.

5. Russian Mathematics and Mysticism

1. Descartes, Meditation III.
2. Arthur Stanley Eddington, *Science and the Unseen World* (New York: Macmillan, 1929), p. 49; Eddington, *New Pathways in Science* (New York: Macmillan, 1935), p. 322.
3. Hermann Weyl, *God and the Universe* (New Haven: Yale University Press, 1932), p. 8.
4. An example, seven hundred pages long and of uneven quality, is Teun Koetsier and Luc Bergmans, eds., *Mathematics and the Divine: A Historical Study* (Amsterdam: Elsevier, 2005).
5. *Istoriko-matematicheskie issledovaniia*, XXXI (1989), p. 147.
6. Plotinus may actually have been Egyptian, but he belonged to the Greek tradition and was a follower of Plato.
7. William James, *The Varieties of Religious Experience* (Cambridge, Mass.: Harvard University Press, 1985), p. 333.
8. Ibid., p. 302.
9. See letter of Luzin to "Petr Afanas'evich" Florensky of May 1, 1906, in S. S. Demidov and A. N. Parshin, eds., "O perepiske N. N. Luzina s P. A. Florenskim," *Istoriko-matematicheskie issledovaniia*, XXI (1989), p. 135.
10. Ibid.
11. See Sviashchennik Pavel Florensky, *Sochinneniia v chetyrekh tomakh*, vol. 3 (1), *Mysl'* (Moscow, 2000), pp. 252–364, and *passim* in other volumes of his collected works.
12. ". . . das Wesen der Mathematik liegt gerade in ihrer Freiheit"; Georg

Cantor, "Über unendliche, lineare Punktmannichfaltigkeiten," *Mathematische Annalen*, vol. 21 (1882), pp. 545–591, on p. 545.

13. The development, under the name "arithmology," of the p-adic numbers by Kurt Hensel in 1897 impressed the Russian Trio.
14. For a modern translation of Memphite theology see Marshall Clagett, *Ancient Egyptian Science*, vol. 1 (Philadelphia: American Philosophical Society, 1989), pp. 305–312, 595–602. We are grateful to John Murdoch for this suggestion. On the Jewish mystical tradition see Gershom Scholem, *Major Trends in Jewish Mysticism* (New York: Schocken, 1995).
15. Henri Lebesgue, "Contribution a l'étude des correspondances de M. Zermelo," *Bulletin de la Société Mathématique de France*, vol. 35 (1907), pp. 227–237, esp. pp. 228, 236. Later, after the intervention of Zermelo and also Richard's letter, the adjective "effective" or "denumerable" would be used, e.g., by Borel: see Hadamard et al., "Cinq lettres sur la théorie des ensembles," *Bulletin de la Société Mathématique de France*, 23 (1905).
16. Roger Cooke, "N. N. Luzin on the Problems of Set Theory," unpublished draft, January 1990, pp. 1–2, 7.
17. A more precise French version would have been "nommer, c'est avoir à faire avec un individu." See the Appendix, p. 207.
18. Letter from Luzin to Arnaud Denjoy, March 4, 1928, in Pierre Dugac, "Nicolas Lusin: Lettres à Arnaud Denjoy, avec introduction et notes," *Archives Internationales d'Histoire des Sciences*," 27 (1977–1978), pp. 179–206.
19. "Récoltes et Semailles, Réflexions et témoignages sur un passé de mathématicien" (1985–86), p. 24; to be published care of IHES. See http://www.fermentmagazine.org/rands/recoltes1.html (accessed on December 30, 2007; translated by Roy Lisker).
20. A. Jackson, "Comme Appelé du Néant," *Notices of the American Mathematical Society*, 51 (no. 10), p. 1197.

6. The Legendary Lusitania

1. Personal communication, S. S. Demidov, as recorded in e-mail from Kantor to Graham, June 29, 2004.
2. M. A. Lavrent'ev, "Nikolai Nikolaevich Luzin," *Russian Mathematical Surveys*, 29, no. 5 (1974), pp. 173–178.

3. E-mail from J.-M. Kantor of June 29, 2004, reporting conversations with Sergei Demidov and oral memoirs of Lazar Lyusternik.
4. Lavrent'ev, "Nikolai Nikolaevich Luzin," *Russian Mathematical Surveys*, 29, no. 5 (1974), pp. 173–178, and *Uspekhi Matematicheskikh Nauk*, 29, no. 5 (1974), pp. 177–182.
5. Esther R. Phillips, "Nicolai Nicolaevich Luzin and the Moscow School of the Theory of Functions," *Historia Mathematica*, 5 (1978), p. 293.
6. For example, the *Dictionary of Scientific Biography* (New York: Scribner, 1970–1990). Egorov is in vol. IV, pp. 287–288; Luzin in vol. VIII, pp. 557–559; Alexandrov in vol. XVII, supplement II, pp. 11–15; Uryson in vol. XIII, pp. 548–549; Stepanov in vol. XIII, pp. 35–36.
7. Lipman Bers, quoted in *Review of U.S.-USSR Interacademy Exchanges and Relations* ("the Kaysen Report"), National Academy of Sciences, Washington, D.C., 1977.
8. L. A. Lyusternik, "The Early Years of the Moscow Mathematical School," *Russian Mathematical Surveys*, 22, no. 4 (1967), p. 60.
9. Ibid., p. 55.
10. Ibid., p. 56.
11. Ibid.
12. "Long live the academy!/ Long live the professors!/ Long live the republic!/ And the one who rules it!/ Long live our city!"
13. V. Stratonov, "Poteria moskovskim universitetom svobody," *Moskovsky universitet, 1755–1930*, Sovremennyia zapiski (Paris, 1930), pp. 199–200.
14. V. V. El'iashevich, A. A. Kizevetter, M. M. Novikov, *Moskovsky universitet, 1755–1930: iubileinyi sbornik*, Izdatelstvo "Sovremennyia zapiski" (Paris, 1930), p. 167.
15. L. A. Lyusternik, "Address at the Jubilee Session of the Moscow Mathematical Society," *Russian Mathematical Surveys*, 20, no. 3 (1965), p. 20. The rhyme is unfortunately lost in translation.
16. El'iashevich et al., *Moskovsky universitet*, pp. 162–163.
17. Lyusternik, "The Early Years of the Moscow Mathematical School," p. 189.
18. Ibid., pp. 171–211.
19. Ibid., pp. 177–178.
20. Ibid., p. 59.
21. Lyusternik, "Address at the Jubilee Session," p. 25.
22. Strategies on attacking the CH problem have appeared recently, such as

the very promising results by W. Hugh Woodin in his "The Continuum Hypothesis, I–II," *Notices of the American Mathematical Society*, 48 (2001), pp. 567–576, 681–690.
23. Waclaw Sierpinski, *Les ensembles projectifs et analytiques* (Paris: Gauthier-Villars, 1950), pp. 44–47. For more information on Suslin, who died tragically young, see V. I. Igoshin, "A Short Biography of Mikhail Yakovlevich Suslin," *Russian Mathematical Surveys*, 51 (3) (1996), pp. 371–183.
24. Ibid.
25. See G. G. Lorentz, "Who Discovered Analytic Sets?" *The Mathematical Intelligencer*, 23, no. 4 (2001), pp. 28–32.
26. S. S. Demidov and B. V. Levshin, eds., *Delo akademika Nikolaia Nikolaevicha Luzina* (St. Petersburg: RKhGI, 1999), *passim*, and especially p. 26.
27. (Unpublished) Hausdorff, Kapsel 61, letter from Alexandrov to Hausdorff, November 29, 1925. This letter will be reprinted in vol. IX of the *Complete Works* of Hausdorff (2010), ed. E. Brieskorn.
28. Fond 496, GARF, State Archive of the Russian Federation.
29. On Shmidt see GARF, State Archive of the Russian Federation, Fond 496. See also Lyusternik, "The Early Years of the Moscow Mathematical School," and P. S. Alexandrov, "Pages from an Autobiography." For a description of Shmidt's connection of Marxism to cosmogony and mathematics, see Loren R. Graham, *Science and Philosophy in the Soviet Union* (New York: Knopf, 1972), pp. 146–156.
30. This and other delightful details of the trip to Petrograd by the Moscow mathematicians are contained in L. A. Lyusternik, "The Early Years of the Moscow Mathematical School," *Russian Mathematical Surveys*, 25, no. 4 (1970), pp. 167–174.

7. Fates of the Russian Trio

1. The biography of the person after whom the church was named, St. Tatiana, seemed relevant to religious believers in Russia in the 1920s. Before the Christianization of the Roman Empire, under the reign of Emperor Alexander Severus (222–235 C.E.), legend says that she had been a secret Christian who was martyred for her faith in the Russian Orthodox Church. She is the patron saint of students, and thus university chapels in the tsarist period were often named after her. In

the Soviet period that followed, those professors and students who secretly continued to practice their faith felt a kinship with St. Tatiana, who had similarly tried to conceal her religion from an oppressive regime.
2. For example, see V. Molodshii, *Effektivizm v Matematike*, Gosudarstvennoe Sotsial'no-Ekonomicheskoe Izdatel'stvo (Moscow, 1938).
3. Cited in L. N. Mitrokhin, "Philosophy of Religion: New Perspectives," *Russian Studies in Philosophy*, 45, no. 3 (Winter 2006–2007), p. 22.
4. René Fülöp-Miller, *The Mind and Face of Bolshevism* (London, 1926).
5. Arnosht (Ernest) Kol'man, *My ne dolzhny byli tak zhit'* (New York: Chalidze Publications, 1982), p. 7.
6. See S. S. Demidov, "Professor Moskovskogo universiteta Dmitrii Fedorovich Egorov i imeslavie v Rossii v pervoi treti XX stoletiia," *Istoriko-matematicheskie issledovaniia*, 39, no. 4 (1999), 138–140.
7. Ibid.
8. Iu. B. Ermolaev, ed., *Nikolai Grigor'evich Chebotaryov* (Kazan, 1994), p. 91.
9. Vitaly Shentalinsky, *Arrested Voices: Resurrecting the Disappeared Writers of the Soviet Regime*, trans. John Crowfoot (New York: Free Press, Martin Kessler Books, 1996), p. 105.
10. Ibid., p. 115.
11. Ibid., pp. 111, 115.
12. S. S. Demidov and B. V. Levshin, eds., *Delo Akademika Nikolaia Nikolaevicha Luzina* (St. Petersburg: RKhGI, 1999).
13. Shentalinsky, *Arrested Voices*, p. 114.
14. Ibid.
15. Ibid., p. 123.
16. Demidov and Levshin, *Delo Akademika Nikolai Nikolaevicha Luzina*, p. 100.
17. Ibid., p. 102.
18. We have been told that during one of his stays in a sanitorium he fathered a daughter by a nurse; the little girl was in bad health, and her fate is unknown. The nurse, however, survived long after Luzin's death in 1950 and was still alive, living in poverty, in the early 1990s, after the end of the Soviet Union. She appealed to the Academy of Sciences for assistance and evidently received some aid, although at this time all institutions in Russia were in dire financial straits.
19. See Sergei S. Demidov and Charles E. Ford, "N. N. Luzin and the Af-

fair of the 'National Fascist Center,'" in *History of Mathematics*, ed. J. Dauben et al. (San Diego: Academic Press, 1995), pp. 137–148.
20. Reprinted from the Archive of the President of the Russian Federation in Demidov and Levshin, *Delo Akademika Nikolai Nikolaevicha Luzina*, p. 18.
21. Loren Graham met Kol'man several times before his death in 1979, both in Moscow and in Cambridge, Massachusetts, and talked about these events with him. The aged Kol'man regretted his actions but did not wish to speak about them in detail. Kol'man and Graham had a mutual friend in the MIT Marxist mathematician and historian of mathematics Dirk Struik, to whom Kol'man dedicated a book interpreting mathematics from the standpoint of Marxism with these words: "To D. J. Struik, as a sign of our fight for a scientific world-outlook." See E. Kol'man, *Predmet i metod sovremennoi matematiki*, Gosudarstvennoe sotsial'no-ekonomicheskoe izdatel'stvo (Moscow, 1936), a gift from Dirk Struik in Loren Graham's possession, and E. Kol'man, *My ne dolzhny byli tak zhit'* (New York: Chalidze Publications, 1982).
22. Kol'man, "Urgent Tasks for Science and Technology and the Role of the Communist Academy," (Moscow-Leningrad: GSI, 1936), pp. 26–40, in Russian quoted by N. S. Ermolaeva, "On the So-Called Leningrad Mathematical Front," *American Mathematical Society Translations* (2), vol. 193 (1999), pp. 261–171.
23. Demidov and Levshin, *Delo Akademika Nikolaia Nikolaevicha Luzina*, p. 18.
24. Loren Graham, "The Socio-Political Roots of Boris Hessen: Soviet Marxism and the History of Science," *Social Studies of Science*, 15, no. 4 (1985), pp. 705–722.
25. Paul Labérenne in *A la lumière du Marxisme* (Paris: Editions Sociales Internationales, 1936).
26. Both Kol'man and Molodshii wrote books in the late 1930s expressing these views and criticizing Luzin, and both indicated in the books that they had expressed such opinions much earlier in lectures. See Kol'man, *Predmet i metod sovremennoi matematiki*, especially pp. 8, 290, and Molodshii, *Effektizm v matematike*, pp. 78–84.
27. Demidov and Levshin, *Delo Akademika Nikolai Nikolaevich Luzina*, p. 257.
28. Ibid., p. 22.
29. Ibid., p. 128.

30. Ibid., pp. 128–129.
31. Letter of Luzin to Shmidt, February 24, 1926, *Istoriko-matematicheskie issledovaniia* (Moscow: Nauka, 1985), p. 279.
32. A. P. Youschkevitch and P. Dugac, "L'affaire de l'Académicien Luzin de 1936," *La Gazette des mathématiciens* (December 1988), p. 34.
33. See Kapitsa's letter in "Pis'mo P. L. Kapitsy V. M. Molotovu 6 Iiulia 1936g," in Demidov and Levshin, *Delo Akademika Nikolai Nikolaevich Luzina*, pp. 261–263.

8. Lusitania and After

1. It should be noted that our chart is not complete because some fields are inadequately represented.
2. Yakov Sinai, *Russian Mathematicians in the 20th Century* (Singapore: World Scientific Publishing Company, 2003); A. A. Bolibruch, Yu. S. Osipov, and Ya. G. Sinai, *Mathematical Events of the Twentieth Century* (Berlin-Heidelberg-New York: Springer, 2006). On the history of the Moscow School of Mathematics, see also Smilka Zdravkovska and Peter L. Duren, eds., *Golden Years of Moscow Mathematics*, 2nd ed. (History of Mathematics, vol. 6), American Mathematical Society and London Mathematical Society, 2007.
3. See the famous letter of Christian Goldbach to Leonhard Euler of June 7, 1742.
4. *www.gay.ru/english/life/religion/florensk.htm*, accessed in 2006.
5. P. S. Alexandrov, "Pages from an Autobiography," *Russian Mathematical Surveys*, 34, no. 6 (1979), pp. 267–302, and 35, no. 3 (1980), pp. 315–358.
6. A. S. Kechris, *Classical Descriptive Set Theory* (New York: Springer-Verlag, 1995), p. 83.
7. P. S. Alexandrov, "Pages from an Autobiography," *Russian Mathematical Surveys*, 35, no. 3 (1980), p. 318.
8. Carlos Sánchez Fernández and Concepción Valdés Castro, *Kolmogórov: El zar del azar*, NIVOLA libros y ediciones, S. L. (Tres Cantos, España, 2003).
9. P. S. Alexandrov, "Pages from an Autobiography," *Russian Mathematical Surveys*, 35, no. 3 (1980), p. 333.
10. Conversations and e-mail messages from Stela Pisareva, director of the museum of Kazan University, and Natalia Zinkina, archivist at the uni-

versity, May 14–15, 2007, as well as a visit by Loren Graham to Kazan University in March 2007.
11. This incident was widely known in the Academy, and was repeated orally many times. One reference to it in print, without the full details, is A. P. Yushkevich, "Encounters with Mathematicians," in Zdravkovska and Duren, eds., *Golden Years of Moscow Mathematics*, p. 24. Another is G. G. Lorentz, "Mathematics and Politics in the Soviet Union from 1928 to 1953," *Journal of Approximation Theory*, 116 (2002), p. 207.

9. The Human in Mathematics, Then and Now

1. F. Patte, "The Karani: How to Use Integers to Make Accurate Calculations on Square Roots," in Gerald G. Emch, R. Sridharan, and M. D. Srinivas, eds., *Contributions to the History of Indian Mathematics* (New Delhi: Hindustan Book Agency, 2005).
2. "In the beginning was the Word"; John 1:1.
3. After the work of the logicians Alfonso Church, Kurt Gödel, and Alan Turing, a precise notion of what an algorithm is emerged around 1935 through the theory of recursive functions. See Martin Davis, ed., *The Undecidable: Basic Papers on Undecidable Propositions, Unsolvable Problems, and Computable Functions* (New York: Raven Press, 1965).
4. A. F. Losev, *Imia: izbrannye raboty, perevody, besedy, issledovaniia, arkhivnye materialy*, ed. A. A. Takho-Godi (St. Petersburg: Aleteiia, 1997); Konstantin Borshch, *Imiaslavie: sbornik bogoslovsko-publitsisticheskikh statei, dokumentov i kommentariev* (Moscow: Pskovaskaia oblastnaia tipografiia, 2003); Elena Gurko, *Bozhestvennaia onomatologiia: imenovanie Boga v imiaslavii, simvolizme i dekonstruktsii* (Minsk: Ekonompress, 2006); Sergei Kremenetsky and Arkhiepiskop Ternopol'sky, *Pravoslavnyi vzgliad no pochitanie imeni Bozhiia: sobytiia na Afone 1913* (L'vov: Izdatel'stvo missionerskogo otdela L'vovskoi eparkhii UPTs, 2003); Tat'iana Senina, *Imiaslavtsy ili imiabozhniki? Spor o prirode imeni Bozhiia i afonsko dvizhenie imiaslavtsev, 1910–1920-kh godov* (St. Petersburg, 2002); D. Leskin, *Spor ob imeni Bozhiem: filosofiia imeni v Rossii v kontekste afonskikh sobitii 1910-kh gg.* (St. Petersburg: Aleteiia, 2004); A. M. Khitrov and O. L. Solomina, *Zabytye stranitsy russkogo imiaslaviia: sbornik dokumentov i publikatsii po afonskim sobytiiam 1910–1913 gg. i dvizheniiu imislaviia v 1910–1918 gg.* (Moscow: Palomnik', 2001); E. S. Polishchuk, *Imiaslavie: Antologiia* (Moscow: Faktorial Press, 2002); N. V.

Skorobogat'ko and A. T. Kazarian, *Imiaslavie: materialy k razresheniiu spora ob Imeni Bozhiem*, Nachala No. 1–4 (Moscow, 1996); Episkop Ilarion (Alfeev), *Sviashchennaia taina tserkvi: vvedenie v istoriiu i problematiku imiaslavskikh sporov*, vol. I (St. Petersburg: Aleteiia, 2002); Episkop Ilarion (Alfeev), *Sviashchennaia taina tserkvi: vvedenie v istoriiu i problematiku imiaslavskikh sporov*, vol. II (St. Petersburg: Aleteiia, 2002); Iu. Rasskazov, *Sekrety imen: ot imiaslavii do filosofii iazyka* (Moscow: Labirint, 2000); Kliment, sviatogorsky monakh, *Imiabozhnichesky bunt, ili plody ucheniia knigi 'na gorakh kavkaza': kratkii istoricheskii ocherk afonskoi smuty* (Moscow: K" Svetu, 2005); Hilarion Alfeyev, *Le Nom grand et glorieux: La vénération du Nom de Dieu et la prière de Jésus dans la tradition orthodoxe*, trans. from Russian by Claire Jounievy, Hiéromoine Alexandre (Siniakov), and Dom André Louf (Paris: Les Éditions du Cerf, 2007); Sviashchennik Pavel Florensky, *Sviashchennoe pereimenovanie: izmenenie imen kak znak peremen v religioznom soznanii*, Khram sviatoi muchenitsy Tatiany (Moscow, 2006).

5. V. M. Tikhomirov, "On Moscow Mathematics—Then and Now," in Smilka Zdravkovska and Peter Duren, eds., *Golden Years of Moscow Mathematics*, American Mathematical Society and London Mathematical Society, 2007, pp. 273–283.

6. See Yiannis Moschovakis, *Descriptive Set Theory* (North-Holland, 1995); Hugh Woodin, "The Continuum Hypothesis," Parts I and II, *Notices of the American Mathematical Society*, 48, nos. 6 and 7 (2001), pp. 567–576 and 681–690.

7. See Claude Dellacherie and Paul-André Meyer, *Probabilities and Potential*, a series of books (Amsterdam-New York: Elsevier North-Holland, 1978–).

8. Pope Benedict XVI, Ratisbonne discourse, September 12, 2006.

9. For Jacques Bouveresse's discussion of Quine's analysis of the existence of real numbers in his lectures at Collège de France in 2007/08, see http://www.college-de-france.fr/default/EN/all/phi_lan/cours_et_seminaires _anterieurs.htm.

10. "How Convincing Is a Proof?" Y. Manin, with discussion by B. H. Neumann and S. Feferman, in *The Mathematical Intelligencer*, 2, no. 1 (1977), pp. 17–24.

11. *Récoltes et Semailles* (R & S), to be published care of IHES. trans. Roy Lisker: www.fermentmagazine.org/rands/recoltes1.html (accessed on December 30, 2007).

12. R & S, p. 24.

13. Review by Oliver Sacks, "In the River of Consciousness," *New York Review of Books*, January 15, 2004; Yu. I. Manin, "Georg Cantor and His Heritage," talk given at the German Mathematical Society and the Cantor Medal award ceremony, September 19, 2002, p. 4; and his reference to S. Dehaene, E. Spelke, P. Pinet, R. Stanescu, and S. Tsivkin, "Sources of Mathematical Thinking: Behavioral and Brain-Imaging Evidence," *Science*, May 7, 1999, vol. 284, pp. 970–974.
14. Personal communication, October 2005.

ACKNOWLEDGMENTS

During the many years we have worked on this book, we have been greatly assisted by many people in several different countries. Although it is impossible to adequately acknowledge all of our intellectual and personal debts, we must mention some people who stand out for their generosity. First of all, we would like to thank Roger Cooke of the University of Vermont, who spent many weeks in Moscow in 1988 and 1989 working on the Luzin papers and then made his voluminous notes available to us (see the Appendix for several of the most interesting findings). In France, Bernard Bru gave us the benefit of his rich knowledge of the history of French mathematics, and followed our research closely.

In Russia, Sergei Demidov and Natalia Ermolaeva answered many of our questions on the history of Russian mathematics, and Alexei Parshin enriched our knowledge of the Name Worshipping movement among Russian mathematicians. Douglas Cameron was a tremendous help in providing us with a number of photographs of members of the Moscow Mathematical School. Barry Mazur, Patrick Dehornoy, and Claude Dellacherie encouraged us to believe that our topic was interesting not only to historians of mathematics but to mathematicians themselves.

The late George Lorentz made his knowledge of the early history of the Moscow Mathematical School available to us. In Paris, Iegor

Acknowledgments

Reznikoff enlightened us on links between our work and Russian music and culture. Pierre Guiraldenq, biographer of Émile Borel, not only helped us on his subject but sent us a photograph. S. S. Kutateladze discussed with us by e-mail the "Luzin Affair" and other topics in the history of Soviet mathematics. Harvey Cox enlightened us on the theological issues in Name Worshipping. Ned Keenan explored with us the interaction of religion and culture in Paris. Michael Gordin gave many helpful comments, including pointing to the interest of the novelist J. D. Salinger in the Jesus Prayer. Karl Hall gave us the benefit of his superb knowledge of the history of physics and mathematics in Russia. Slava Gerovitch has been an intellectual stimulus for many years. John Murdoch noted the importance of "naming" in medieval Europe and in ancient Egyptian culture. Donald Fanger illustrated the links between symbolism in mathematics and in Russian culture and literature. Peter Buck discussed with us the history of both mathematics and the social sciences. Masha Vorontsova and Elena Lyapunova assisted in a number of ways, including finding information on Dmitri Egorov. Mikhail Shemiakin, a relative of Egorov, gave us a photograph of Anna Egorova and described the Egorov and Grzhimali families to us. Cliff Erickson, a physicist with an unusual knowledge of Russia, read a version of the manuscript and made helpful suggestions. Peter Galison encouraged us to combine deep scholarship with clarity and readability.

Sol and Laurie Garfunkel were generous with their time, with hospitality, and with friendship. Victor Guillemin described his stays at the Hotel Parisiana in Paris and the recollections of the Chamont daughters of Egorov and Luzin. Nancy Holyoke helped us improve the style, readability, and organization of the book. Jonathan Dickinson displayed great interest in our research and assisted in English-language translations from French. Egbert Brieskorn discussed with us the German elements of the story, especially Hausdorff. Dmitri Baiuk in Moscow attempted to find a remaining important and elusive source. Tatiana Yankelevich, Sarah Failla, Donna Griesenbeck, and Helen Repina at the Davis Center for Russian and Eurasian

Studies at Harvard University helped in many different ways in finding sources and translations. Alberto Arabia assisted on computer and technological matters. Stela Pisareva, director of the museum of Kazan University, and Natalia Zinkina, archivist at Kazan, helped us to find sources describing the fate of Egorov there.

Kathleen McDermott, our editor at Harvard University Press, came to a talk that Loren Graham gave on Name Worshipping and subsequently promoted our book at the Press and with her fellow editors. We appreciate her support and advice. Mary Ellen Geer, Senior Editor at the Press, did a superb job of improving the style and composition of the book.

On a more personal level, we would like to thank Sheila Biddle, a true friend for decades. Finally, and most fervently, we would like to express our appreciation to members of our families, including Meg Graham, a fine literary critic, and our wives, Patricia and Dominique, who displayed striking intelligence, patience, and good cheer as we journeyed to libraries and archives in the United States, France, and Russia.

Needless to say, none of these good people bear any responsibility for the errors and misjudgments that may remain in the book.

INDEX

Abkhazia, 13
Academy of Sciences of the Soviet Union, 138, 186, 197
Adelphopoesis (brother-making), 84
Aikenval'd, T. Iu., 104
Alexandra, tsarina, 10, 14
Alexandrov, Pavel, 49, 63, 103, 116, 117; and A-sets, 119, 120; as critic of Luzin, 121, 122; opposition to DST, 149; opposition to Luzin, 150, 153, 154, 157, 172; and election to Academy of Sciences, 161, 164, 166; and Uryson, 174–178; photo with Brouwer and Uryson, 176; death of Uryson, 177–178; and Kolmogorov, 179–180; photo with Kolmogorov, 185
Anaximander of Miletus, 21
Andreevsky Monastery, Mt. Athos, 11
Anti-Name Worshippers, on Mt. Athos, 9
Anti-Semitism, 45, 196
Apeiron, 21, 22, 23

Appell, Marguerite (Camille Marbo), 37, 41, 45
Appell, Paul, 29, 31, 41, 42, 47
Arabia, Alberto, 230
Archimedes, 41
Aristotle, 22, 23, 148, 188
Arithmology, 67, 71
Arnol'd, V. I., 16
A-sets, 119, 210
Axiomatic method, 63
Axiom of Choice, 50, 56; French discussion of, 56–61, 199, 206, 207, 210

Baire, René, 24, 31, 35, 44, 45, 48, 49, 50–54, 205; psychological difficulties of, 53, 64; suicide of, 63; as grandfather of DST, 120; mentioned in Luzin trial, 155
Baire space, 50
Baiuk, Dmitri, 229
Bakh, A. N., 153
Bari, Nina, 104, 116, 153, 154, 155, 164, 166; death of, 195
Belhoste, B., 214n4

{ 231 }

Bely, Andrei, 67, 75, 79, 87, 91, 97
Bendixson, Ivar Otto, 117
Bergmans, Luc, 218n4
Beria, Lavrentii, 158
Bernoulli, Jacob, 69
Bernshtein, S. N., 153
Bernstein, S. A., 116
Bers, Lipman, 220n7
Bertrand, Joseph, 41, 42, 62
Beskin, N. M., 76
Bibliothèque Sainte-Geneviève, 81
Biddle, Sheila, 230
Boileau, Nicolas, 36
Bois-Reymond, Paul du, 29–30
Bolibruch, A. A., 224n2
Bolzano, Bernard, 25
Boole, George, 199
Borel, Émile, 24, 31, 35, 37–44, 206; marriage to Marguerite Appell, 42, 45, 47, 49, 51; meets Russell, 55; and Axiom of Choice, 58; differences with Lebesgue, 61–64; reservations about set theory, 63; and science in WWI, 64; and Dreyfus Affair, 73, 80, 99; and B-sets, 117; as grandfather of DST, 120; mentioned in Luzin trial, 155–156; protests persecution of Luzin, 157, 175; quarrel with Lebesgue, 186
Borel, Honoré, 38
Borges, Jorge Luis, 19
Bourbaki group, 60, 199, 200
Boutroux, Émile, 34, 42
Boutroux, Pierre, 42
Bouveresse, Jacques, 199
Brahmans and divine numbers, 190

Brieskorn, Egbert, 221n27, 229
Brouwer, L. E. J., 175, 176, 178–179, 199
Bru, Bernard, 214n11, 228
B-sets (Borelian sets), 41, 117, 119
Buck, Peter, 229
Buckle, Henry Thomas, 66
Bugaev, Nikolai V., 67, 72, 79, 89
Bukharin, Nikolai, 115, 127, 131, 148
Bulatovich, Alexander (monk Antony), 12, 16
Bulgakov, Sergei, 17
Burali-Forti, Cesare, 55

Cameron, Douglas, 228
Cantor, Georg, 20, 24; and set theory, 25, 26; and Cantor ternary set, 27, 28, 35, 49; and paradoxes, 55, 88, 96, 116–117, 199, 210
Cardinal number, 27, 55, 95
Catacomb Church, 133–134, 136
Cauchy, Augustin, 39
Chaliapin, Feodor, 74
Chamont, M., 82
Chebotaryov, Nikolai, 131–133, 137–139, 184, 196–197
Chebyshev, Pafnuty, 121
Choquet, Gustave, 198, 211n1
Church, Alonzo, 198, 209, 225n3
Church of St. Tatiana the Martyr, 3, 4, 82, 84, 105, 110–111, 126
Clemenceau, Georges, 41
Cohen, Paul, 61, 121, 206, 209, 211n2
Communist Party, France, 157
Comte, Auguste, 35, 37
Concepción, Valdés Castro, 224n8

Index

Continuum/Continuum Hypothesis (CH), 26, 27, 30, 31, 35, 56, 63, 88, 116, 167, 172, 198, 207, 209
Cooke, Roger, 205, 211n3, 219n16, 228
Corsi, P., 213n9
Courant, Richard, 175
Couturat, Louis, 50
Cox, Harvey, 229
Curie, Marie, 43
Curie, Pierre, 43

Darboux, Jean Gaston, 31, 39, 41, 47, 50, 53, 71, 72
Darwin, Charles, 88
Dauben, Joseph, 223n19
David, archimandrite, 17
Davis, Martin, 225n3
Dedekind, Richard, 26, 28
Dehornoy, Patrick, 228
Dellacherie, Claude, 226n7, 228
Demidov, Sergei, 72, 216nn1,5,7,8, 217nn8,9,14, 218nn9,1, 220n3, 221n26, 222nn6,12,19, 223nn23,27, 224n33, 228
Denjoy, Arnaud, 38, 53, 54, 98, 155–156, 198, 207; mentioned in Luzin trial, 155–156
Denjoy, Fabrice, 155
Denjoy, René, 99
Derrida, Jacques, 18
Desanti, Jean-Toussaint, 199
Descartes, René, 35
Descriptive Set Theory (DST), 117; birth of, 118, 198
Dickinson, Jonathan, 229
Dieudonné, Jean, 63
Dini, Ulisse, 51
Dreyfus Affair, 45–47, 73

Dugac, Pierre, 211n4, 213nn11,12, 214n11, 215n17, 219n18, 224n32
Dummett, Michael, 199
Duncan, Isadora, 173
Duren, Peter, 224n2, 225n5

École Normale Supérieure, 51, 81
École Polytechnique, 51
Eddington, Arthur Stanley, 92
Église Sante-Geneviève, 81
Egorov, Dmitri, 16, 17, 18, 19, 31; and Dreyfus Affair, 45, 66, 67, 71–77, 79, 80; entertains students at home, 102; standing among world mathematicians, 103; style of teaching, 108, 109; pictured with Luzin and Sierpinski, 120; after Russian Revolution, 125–127; and Chebotaryov, 132; arrest of, 136; in exile, 136; death of, 137–138; founder of Moscow School of Mathematics, 162–163, 173
Egorov surfaces, 71
Eiges, Aleksandr, 171
Eiges, Ekaterina, 173, 174
Erickson, Cliff, 229
Ermolaev, Iu. B., 222n8
Ermolaeva, N. S., 223n22, 228
Ern, V. F., 86
Esenin, Sergei, 173
Eudoxus, 35, 41
Euler, Leonhard, 53

Failla, Sarah, 229
Fanger, Donald, 229
Feferman, S., 226n10
Fernández, Carlos Sánchez, 224n8
Ferreirós, José, 213n1

{ 233 }

Fersman, A. E., 153
Fligye, Irina, 145
Florenskaia, Anna Mikhailovna, 140
Florensky, Pavel, 15, 17, 24, 66; and renaming, 68, 75, 76, 79; *The Pillar and Foundation of Truth*, 83, 169; *Holy Renaming*, 83–84, 86–90, 95, 96, 125, 140; second arrest of, 140; third arrest and confession of, 142; death of, 145
Fock, Vladimir, 158
Ford, C. F., 216nn4,5, 217nn8,11,12,14, 218n17, 222n19
Fourier, Joseph, 30, 35
Fréchet, Maurice, 31, 170
Free Will, 67–71
Frege, Gottlob, 200
French Revolution, 81
Friedman, Harvey, 213n6
Frobenius, Ferdinand George, 72
Fülöp-Miller, René, 16
Functions, continuous and discontinuous, 40; discontinuous of class one, two, 48, 50, 52–53, 59; semi-continuous, 53, 67, 68, 71, 84, 87–89, 97

Galileo, 24
Galison, Peter, 229
Garfunkel, Sol and Laurie, 229
Gaunilon, monk of Noirmoutiers, 59
Gel'fand, I. M., 164, 194
Gérard, E., 45
Gerovitch, Slava, 229
Gippius, Zinaida, 169
Glossalia, 12
God-Builders, 127

Gödel, Kurt, 61, 121, 206, 225n3
Gorbunov, N. P., 153
Gordin, Michael, 229
Gorky, Maxim, 141
Graham, Meg, 230
Graham, Patricia Albjerg, 230
Grave, Dmitri, 132
Gregory of Rimini, 24
Griesenbeck, Donna, 229
Grigory Palama, 12–13
Grothendieck, Alexander, 92, 100, 199–201
Grzhimali, Anna (Aida), 74–75, 108
Grzhimali, Ivan, 74–75, 108
Grzhimali, Natalya, 74–75, 108
Guillemin, Victor, 82, 229
Guiraldenq, Pierre, 214n5, 229
Gurko, Elena, 225n4
Gustafson, Richard, 169

Hadamard, Jacques, 31, 45–46, 51, 72, 80, 157
Hall, Karl, 229
Haney, Frank, 217n13
Hausdorff, Felix, 49, 117, 170, 175
Heine-Borel Theorem, 40
Hermite, Charles, 29, 30, 39, 42, 45, 50, 51
Hessen, Boris, 148, 223n24
Hesychast monks, 12
Hilbert, David, 28, 30, 31, 33, 34, 35, 56, 57, 63, 72, 116, 170, 175, 199, 211
Hilbert School, 60, 200
Hitler, Adolf, 150, 160
Holyoke, Nancy, 229
Holy Synod, 9, 16
Homosexuality and Lusitania, 168–170, 185–186

Index

Hoover, Herbert, 109
Hopf, Heinz, 181
Hume, David, 192

Ibsen, Henrik, 173
Idealism, 115
Igoshin, V. I., 221n23
Ilarion, monk, 12, 13, 14
Ineffability, 21, 95, 100
Institute of Mathematics and Mechanics, Moscow University, 135
Ioakim III, patriarch, 14
Italian school of mathematics, 31

Jackson, A., 219n20
James, William, 94–95
Jesus Prayer, 8, 9, 12, 13–14, 16, 133, 136
Jordan, Camille, 40, 50

Kant, Immanuel, 35, 55, 210
Kantor, Dominique, 230
Kapitsa, Peter, 157–160
Kataev, Nikolai, 180
Kazan, city and university, 132, 136
Kechris, A. S., 224n6
Keenan, Ned, 229
Keldysh, Ludmila, 59, 164, 165
Keldysh, M. V., 164
Khinchin, A. Ia., 145, 153, 154, 166
Khitrov, A. M., 225n4
Khrisanf, Father, 14–15
Khrushchev, Nikita, 158
Kleene, S. C., 198
Klein, Felix, 40, 72
Kliuev, Nikolai, 173
Koetsier, Teun, 218n4
Kol'man, Ernst, 129; critic of Egorov, Florensky and Luzin, 129–131, 134–135, 146–148, 150; sets trap for Luzin, 151–153; denunciation of Luzin, 156, 157
Kolmogorov, Andrei, 103, 122, 154, 164, 166, 180–187; relationship with P. S. Alexandrov, 170, 180–187; relationship with P. S. Uryson, 170, 171; and Luzin, 180; and Stepanov, 180; and Luzin trial, 183; photo with Alexandrov, 185
Komarovka, 182
König, Julius, 56
Kostitsin, V. A., 78
Kronecker, Leopold, 28, 29, 40
Krzhizhanovsky, Gleb, 153, 158–160
Kurosh, A. G., 166
Kutateladze, S. S., 229
Kuznetsov, P. I., 217n8

Labérenne, Paul, 148, 223n25
Landau, Lev, 158
Langevin, Paul, 42, 43, 63, 157
Laplace, Pierre-Simon, 30, 35
Laub, Ferdinand, 74
Lavrent'ev, M. V., 164, 166, 217n8
Law of Large Numbers, 69
Lazarev, P. P., 123
Lebesgue, Henri, 24, 31, 35, 40, 42, 44–50, 207; and Axiom of Choice, 56, 58; and Banach, 59; differences with Borel, 61–64, 72; and Dreyfus Affair, 73, 80; and naming, 98; and Lusitania, 116, 118; his mistake and Lusitania, 119; as grandfather of DST, 120; criticism by Kol'man, 148;

{ 235 }

Lebesgue, Henri *(continued)*
    mentioned in Luzin trial, 155,
    175; quarrels, 186
Lebesgue Integral, 47, 50
Lejeune-Dirichlet, Peter Gustav, 53
Lenin, Vladimir, 125, 128
Leskin, D., 225n4
Levshin, B. V., 216n8, 221n26,
    222n12, 223nn23,27, 224n33
Lisker, Roy, 226n11
Lobachevsky, Nikolai, 138, 171, 184
Lorentz, G. G., 221n25, 228
Losev, Aleksei, 17, 225n4
Lubianka prison, 112, 140, 144
Lunacharsky, Anatoly, 113, 127, 131
Lusitania, 101–124; lasting influence, 194
Luzin, Nikolai, 19; and Plotinus, 24, 31, 34; arrival in Paris, 50, 58, 66, 77–86; psychological crisis of, 79–83; and naming, 91, 98, 99; and mysticism, 93–96; and William James, 94–95; teaching of, 105–107, 118, 171, 193; after Russian Revolution, 125–126; and the "nationalist-fascist center," 142–143; criticism by Kol'man, 148; criticized by his students, 149–150; with Denjoy family, 155; Luzin Affair and narrow escape, 159–160, 187; and Shnirel'man, 162
Luzina, Nadezhda Mikhailovna, 84, 85, 93, 107, 155
Lyapunova, Elena, 229
Lycée Henri IV, 51
Lycée Lakanal, 50
Lyell, Charles, 88
Lysenko, Trofim, 185

Lyusternik, L. A., 101, 103, 113, 115, 154, 166, 167
Lyusternik-Schnirelmann Category, 167

Machine-Worshippers, 128
Malenkov, Georgi, 158
Malygin, Mikhail, 78
Malygina, Nadezhda, 79
Manin, Y., 226n10, 227n13
Marbo, Camille, 37, 43, 47, 62, 63, 64
Markov, A. A., 68–71, 121, 197
Markov chains, 71, 197
Martens, Ludwig, 141, 143
Marxism, 67, 87; and mathematics, 147
Mathematics Department, Moscow University, 162–163
Maurey, Bernard, 214n7
Mazur, Barry, 228
Mekhlis, L. Z., 152
Men'shov, Dmitri, 124, 166
Merezhkovsky, Dmitri, 169
Meyer, Paul-André, 226n7
Minkowski, Hermann, 34, 72
Mitrokhin, L. N., 222n3
Mittag-Leffler, Gösta, 29, 30
Mlodzeevsky, B. K., 72
Molodshii, Vladimir Nikolaevich, 126–127, 148
Molotov, Viacheslav, 158
Mongré, Paul, 117
Montel, Paul, 42
Moore, G. H., 213n7
Morozov, V. V., 138, 184
Moschovakis, Yiannis, 226n6
Moscow Mathematical Society, 16, 67, 87, 135, 184

Index

Moscow School of Mathematics, 105, 115, 148, 162–163, 187, 192
Mt. Athos, 7, 8–11, 96
Murdoch, John, 219n14, 229
Musil, Robert, 117

Name Worshippers, on Mt. Athos, 9
Naming, named *(nommé, imennoe)*, 49, 50, 58, 61, 62, 68, 84, 97, 98, 99, 206; by Lusitanians, 123, 191, 195
"Nationalist-fascist center," 142–143, 146
Nekrasov, P. A., 68–71, 121, 197
Neoplatonism, 94
Neumann, B. H., 226n10
New Economic Policy, 127
Newenglozski, M., 39
Newton, Isaac, 157, 197
Nicholas of Cusa, 24–25
Nicholas II, tsar, 7, 10, 15
Nikon, archbishop of Vologda, 11
Noether, Emmy, 175
Normal numbers, 58
Novikov, M., 109, 111, 112, 115
Novikov, P. S., 164, 166
Novosyolov, M. A., 96

Ordinal numbers, 27
Osipov, Yu. S., 224n2
Ottoman Empire, 8

Painlevé, Paul, 41
Pantaleimon Monastery, Mt. Athos, 9, 12, 15
Pantheon, 81
Parisiana, Hotel, 81
Parshin, A. N., 217nn8,9, 218n9, 228

Pascal, Blaise, 92, 197
Patte, F., 225n1
Peano, Giuseppe, 51
Perrin, Jean, 42
Peshkova, Yekaterina, 141
Pevzner, B. I., 104
Phillips, E. R., 217n8, 220n5
Picard, Émile, 29, 31, 35, 42, 45, 51, 59, 60
Pisareva, Stela, 224n10, 230
Planck, Max, 56
Plato, 26, 148, 188, 190
Platonism, 115
Plotinus, 23, 93–94
Poincaré, Henri, 30, 31, 34, 41, 46, 47, 50, 51; and Richard's Paradox, 56, 72; and Dreyfus Affair, 73, 170
Poincaré, Raymond, 46
Polikarpov, K., 145
Polishchuk, E. S., 225n4
Polovinkin, S. M., 216n2, 217nn8,9
Pólya, George, 63
Pontriagin, Lev, 154, 164, 166
Potemkin, V. P., 157
Prayer of the Heart, 12
Pure Russian Orthodox Church, 134, 136
Pythagoras, 21, 22, 92, 200

Quetelet, L. A. J., 66, 69
Quine, W. V. O., 199

Rachmaninoff, Sergei, 74
Raspail, François-Vincent, 42
Rasputin, Grigori, 14
Reform of French education (1902), 62
Repin, Ilya, 74

Repina, Helen, 229
Revolution of 1905, Russia, 80
Reznikoff, Iegor, 228–229
Richard, Jules/ Richard's Paradox, 55–56, 56, 206
Roberval, Gilles de, 45
Rockefeller Foundation, 129
Rozanov, Vasilii, 169
Rozhanskaia, I. A., 104
Russell, Bertrand, 23, 50, 55, 200
Russian Psychological Society, 67

Sabler, V. K., 10
Sacks, Oliver, 226n13
Saint Anselme, 59
Salinger, J. D., 13
Second International Congress of the History of Science (London, 1931), 148
Seignobos, Charles, 42
Sergiev Posad, 82, 90, 125, 127, 140, 141
Sergius, metropolitan, 134
Shakhty Trial, 135
Shemiakin, Mikhail, 229
Shentalinsky, Vitaly, 222n9
Shields, A. L., 216n4, 217n8
Shikman, A. P., 217n13
Shipulinsky, Z. A., 7, 10
Shmidt, Otto, 122–123, 128–129, 135, 153, 155
Shmidt, Pyotr, 90
Shnirel'man, Lev, 103, 113, 153, 162, 166–168; death of, 195
Sierpinski, Waclaw, 118, 119, 164, 198, 207
Sinai, Ya. K., 164, 224n2
Smirnitskaia, Maria, 132, 137–138, 196–197

Sobolev, S. L., 153, 154, 164
Solomina, O. I., 225n4
Solovetsk Prison Camp, 143–144
Solzhenitsyn, Alexander, 186
Sorbonne, 81
Spinoza, 28
St. Affrique, 38
St. Petersburg School of Mathematics, 69
St. Tatiana, 133, 221n1
Stalin, Joseph, 152, 158–160
Steklov Institute of Mathematics, 182, 183
Stepanov, Viacheslav, 103, 180
Stratonov, V., 112
Struik, Dirk, 223n21
Suslin, Mikhail, 62, 118, 119, 120, 166, 210

Tacchi, Jean-Pierre, 214n7
Tannery, Jules, 42
Tannery, Paul, 42
Tchaikovsky, Peter, 74
Tikhomirov, V. M., 193, 226n5
Tikhon, patriarch, 17
Tikhonov, Andrei, 166
Tolstoy, Leo, 73, 190
Transfinite numbers, 49, 59, 105, 117, 148
Trinity–St. Sergey Monastery, 140
Trotsky, Leon, 125, 126
True Church Movement, 126
Tukhachevsky, marshall, 160
Turing, Alan, 198, 225n3

Uryson, Pavel, 103, 116, 118, 124, 166, 171, 174; and Alexandrov, 174–179; photo with Alexandrov and Brouwer, 176

Index

Valery, Paul, 63
Vernadsky, Vladimir, 73
Vinogradov, I. M., 153
Volkov, A. A., 112
Volterra, Vito, 51, 62
Vorontsova, Masha, 229

Weierstrass, Karl, 39
Weil, André, 157
Weyl, Hermann, 19, 20, 21, 22, 92
Woodin, W. Hugh, 221n22, 226n6

Yagoda, Genrikh, 140
Yankelevich, Tatiana, 229
Yessenin-Volpin, Alexander, 23
Youschkevitch, A. P., 224n32

Zdarvkovska, Smilka, 224n2, 225n5
Zeno, 22, 23
Zermelo, Ernst, 56, 210
Zermelo-Fraenkel axioms, 198
Zinkina, Natalia, 224n10, 230